# HOW OLD IS THE EARTH?

More than twenty years ago PATRICK M. HURLEY read Arthur Holmes' little book *The Age of the Earth* and since has been unable to tear himself away from the subject. Geological age measurement and the application of nuclear physics to geology have been his special scientific interests, and *How Old Is the Earth?*, his first book, is a distillation of what he has learned about these fascinating subjects in his unusually varied career.

Born in Hong Kong, in 1912, he moved with his family to Vancouver Island, British Columbia, at the age of nine. He was graduated from the University of British Columbia with a B.A. degree and a B.A.Sc. degree in engineering. After prospecting for gold for three years, he went to M.I.T. on a fellowship from the Royal Society of Canada and carried on research in radioactivity and geologic age measurement. He received his Ph.D. in 1940.

In World War II, Dr. Hurley worked with the United States Navy, under the National Defense Research Committee, on anti-submarine warfare and underwater ballistics. A year of research in geophysics at the University of Wisconsin followed, and in 1946 he returned to M.I.T., where he now is Professor of Geology and Executive Officer of the Geology Department. Besides his academic work, he has explored for minerals from the Arctic Circle to the Equator, fulfilled research contracts for the Office of Naval Research and the Atomic Energy Commission, and contributed articles to various technical publications, to *Science* magazine, the *American Journal of Science,* and the *Scientific American.*

Dr. Hurley is keenly interested in the art of teaching and in educational policy. He has served on many national scientific committees and for the last several years on M.I.T.'s principal educational policy committees. It is his conviction that today's educational program needs to emphasize a "broader understanding of our physical environment and a reaffirmation of human values." He takes particular enjoyment in working in the laboratory with students exploring earth history, and he hopes *How Old Is the Earth?* will "start some new explorers in earth science."

Dr. Hurley, his wife, and their three children live in Lexington, Massachusetts.

# HOW OLD IS THE EARTH?

Patrick M. Hurley

Published by
Anchor Books
Doubleday & Company, Inc.
Garden City, New York
1959

Available to secondary school
students and teachers through
Wesleyan University Press Incorporated
Columbus 16, Ohio

COVER DESIGN BY GEORGE GIUSTI
ILLUSTRATIONS BY R. PAUL LARKIN
TYPOGRAPHY BY EDWARD GOREY

Figures 21, 22, 23, 25 have been adapted from ones appearing originally in the *Scientific American*.

Library of Congress Catalog Card Number
59–11599

# The Science Study Series

The Science Study Series offers to students and to the general public the writing of distinguished authors on the most stirring and fundamental topics of physics, from the smallest known particles to the whole universe. Some of the books tell of the role of physics in the world of man, his technology and civilization. Others are biographical in nature, telling the fascinating stories of the great discoverers and their discoveries. All the authors have been selected both for expertness in the fields they discuss and for ability to communicate their special knowledge and their own views in an interesting way. The primary purpose of these books is to provide a survey of physics within the grasp of the young student or the layman. Many of the books, it is hoped, will encourage the reader to make his own investigations of natural phenomena.

These books are published as part of a fresh approach to the teaching and study of physics. At the Massachusetts Institute of Technology during

1956 a group of physicists, high school teachers, journalists, apparatus designers, film producers, and other specialists organized the Physical Science Study Committee, now operating as a part of Educational Services Incorporated, Watertown, Massachusetts. They pooled their knowledge and experience toward the design and creation of aids to the learning of physics. Initially their effort was supported by the National Science Foundation, which has continued to aid the program. The Ford Foundation, the Fund for the Advancement of Education, and the Alfred P. Sloan Foundation have also given support. The Committee is creating a textbook, an extensive film series, a laboratory guide, especially designed apparatus, and a teacher's source book for a new integrated secondary school physics program which is undergoing continuous evaluation with secondary school teachers.

The Series is guided by the Board of Editors of the Physical Science Study Committee, consisting of Paul F. Brandwein, the Conservation Foundation and Harcourt, Brace and Company; John H. Durston, Educational Services Incorporated; Francis L. Friedman, Massachusetts Institute of Technology; Samuel A. Goudsmit, Brookhaven National Laboratory; Bruce F. Kingsbury, Educational Services Incorporated; Philippe LeCorbeiller, Harvard University; Gerard Piel, *Scientific American;* and Herbert S. Zim, Simon and Schuster, Inc.

# CONTENTS

# Introduction

"Before the hills in order stood,
Or earth received her frame . . ."
ISAAC WATTS (1674–1748)

Ever since man has had the ability to reason, he has speculated upon the universe about him. The nature of the earth and stars and their origins have been described ten thousand times, in different tongues, in prehistory and history. Each of these descriptions has rested on some evidence, some witnessed phenomenon. But only since the seventeenth-century renaissance in science has there been a conscientious effort to seek explanations for all phenomena within the bounds of a minimum number of physical laws, to eliminate superstition and dogma from the realm of science.

This extension of the disciplines of reasoning to an explanation of our earth, the solar system, the galaxies, and the universe has led to present-day

11

hypotheses that only new evidence—more powerful applications of physical principles or a more comprehensive fitting of them to the myriad observations of fact—can challenge. Any theory will be modified as time goes by. But lest this present history of the earth be dismissed as visionary, let us remember that to be accepted, even temporarily, a theory must have been stripped of fancy and defended before the scientific world of the day.

The development of scientific thought on the origin of the earth is a particularly constructive record for the student of science: it illustrates the struggle to bring together diverse and conflicting hypotheses and observations and shows how widely opinion can swing in a few years. Indeed, within the lifetime of the youngest reader of this book new knowledge of the atomic nucleus has altered radically our picture of the universe.

The measurement of time by study of the continuous breakdown of radioactive elements has had great impact on science and philosophy. We have learned that the naturally occurring radioactive elements are constantly decreasing in abundance, and this phenomenon forces upon us a new realization. It demands a creation of these elements, and therefore probably of all elements, at some definite time in the not-too-distant past. The elements of the world we live in definitely were not in existence forever; therefore, neither was this earth, nor this solar system, nor our galaxy of stars.

What was the nature of this creation? When did it occur, and how did we reach our present state of being? These are questions for the mature intellect. Studying these processes does not diminish their grandeur. The more clearly we see it, the genesis of this part of the universe, from the birth of the stars to the evolution of the human mind, increases in its inspiring majesty.

Our earth has had a dynamic history, not only in the development of human forms but also in the great changes that have brought to a cold, desolate waste like the moon our present hospitable climates, the lands and seas, the wealth of minerals and fuels, the mountains and plains. What great forces generated these sweeping changes? Mountains are short-lived objects on the earth's surface. Unlike people, they are largest when youngest. Under the eroding actions of ice and rain, they mature to low, rounded forms, and in a time that is short in the earth's history they wear to flat plains, leaving no evidence of their existence except in the rocks and structures of their roots. But as mountains have worn away, other geologic features have appeared. What is the source of the vast energy that develops belts of volcanoes pouring forth molten lava and shaking the ground with earthquakes, that lifts continents above the ocean floor and makes the earth's surface buckle and wrench? What energy concentrates copper and gold into deposits that glitter under the miner's lamp like a jewelry store?

Now we recognize the answer. It is the energy of the radioactive breakdown of nuclei of atoms like uranium and thorium. This occurs in small amounts in all the ground we stand on. The energy released is transformed into heat which, as it has for billions of years, flows to the surface. When the shift of surface material, such as deposition of sediment at the edge of a continent, blocks this flow, the temperature beneath builds up to the point where rocks melt, and the crust becomes weak enough to buckle.

Not only has radioactivity supplied most of the energy for the earth's great geologic events; it also measures the time at which these events have occurred. As we shall see in later chapters, each grain of sand, each minute crystal in the rocks about us is a tiny clock, ticking off the years since it was formed. It is not always easy to read them, and we need complex instruments to do it, but they are true clocks or chronometers. The story they tell numbers the pages of earth history.

But let us start at the beginning. . . .

# CHAPTER I

# The Structure of the Earth

The earth is almost a sphere, but the centrifugal force of its rotation causes a bulging at the equator and a slight flattening at the poles. Its radius is about 6400 kilometers (or 3960 miles). We know that the earth is divided into distinct layers, and we believe that it is composed of two main types of material.

If you could see the earth in cross section, you would find a rather sharp division between the *core,* or central part, and the *mantle,* or outer part. This is shown in Fig. 1. As you can see in the diagram, the core itself appears to be divided into inner and outer parts. There are several lines of evidence which lead to the belief that the core is composed 90 per cent of metallic iron and possibly silicon, in the proportion of three parts to two, and 10 per cent of nickel, largely in a fluid state. The inner core appears to have the properties of a solid.

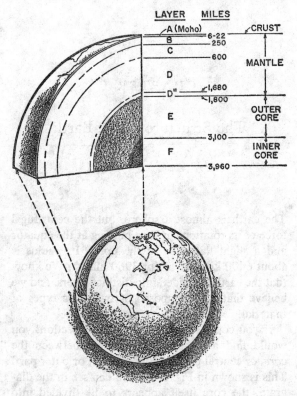

*Fig. 1. The earth's surface is well known to us, but to explain the nature of its interior, we still have only hypotheses, and they conflict. The deepest hole penetrates only about 5 miles. Information about the interior is derived from seismic wave velocities, meteorites, volcanic eruptions, and other measured and observed phenomena.*

16

The existence of a sharp boundary between core and mantle is inferred from observed changes in the direction and speed of shock, or *seismic,* waves passing through the earth. These changes, which can be measured with an instrument called the *seismograph,* indicate differences in the physical properties of the core and mantle. On the evidence of the seismograph, R. D. Oldham of England established in 1906 that the core exists. In 1914, Beno Gutenburg of Germany, later a professor at the California Institute of Technology, showed that the separation between core and mantle occurs about halfway between the surface and the center of the earth. This boundary is called the *Wiechert-Gutenburg discontinuity.* (Ernst Wiechert of Germany contributed to the development of the seismograph.)

The mantle is believed to be composed predominantly of oxygen, magnesium, and silicon (with a little iron) in a ratio of four atoms of oxygen to two atoms of magnesium and one of silicon; all the other elements combined equal only about one tenth the total amount of these four. If a piece of the mantle were brought to the surface, melted, and allowed to cool slowly, it would look like dark green rock, and it would be made up largely of the mineral olivine, magnesium iron silicate. Because of the high pressure and temperatures existing in the mantle (white hot in the lower part), it is not known in which principal mineral forms these elements occur, but it *is* known that they are in the solid state.

The outer surface of the mantle has a very shallow skin layer of different composition, known as the *crust*. One of the sharp changes in seismic wave velocity I have mentioned marks the boundary between the crust and the rest of the mantle. A Croatian seismologist, A. Mohorovicic, discovered it in 1909 while studying the seismograph of an earthquake in the Balkans, and it gets its name from him, the *Mohorovicic discontinuity,* or *Moho.* We find the Moho at different depths in different regions but generally at about 35 kilometers (22 miles) under the continents and 10 kilometers under the oceans.

The composition of the crust is quite varied; it ranges from light-colored rocks like granite, rich in silica, alumina, and alkalies, to dark-colored rocks like those of the Hawaiian Islands, rich in iron oxide and magnesia. A thin layer of sediments, which have turned into rocks like shale, sandstone, or limestone or, if buried deep enough, have been baked into more crystalline rocks, covers a large extent of the crust surface. These rocks of sedimentary origin make up only a small fraction of the crust's total mass. The surface of the earth also has masses of water, referred to as the *hydrosphere,* and an envelope of gases, called the *atmosphere*.

The hydrosphere and atmosphere aside, eight elements predominate in the composition of the earth's crust; they make up about 99 per cent of its mass. In terms of numbers of atoms, oxygen accounts for more than 60 per cent of the total.

THE COMMONER ELEMENTS IN THE EARTH'S CRUST
COMPARED WITH AVERAGE STONY (CHONDRITIC) METEORITE

| Element | Atomic radius (×10⁸ cm) | Crust | | | Meteorite | |
|---|---|---|---|---|---|---|
| | | Weight (Per cent) | Atom (Per cent) | Volume (Per cent) | Atom (Per cent) | |
| Oxygen | 1.32 | 46.60 | 62.55 | 91.97 | 58.6 |
| Silicon | 0.39 | 27.72 | 21.22 | 0.80 | 16.7 |
| Aluminum | 0.57 | 8.13 | 6.47 | 0.77 | 1.5 |
| Iron | 0.82 | 5.00 | 1.92 | 0.68 | 6.3 |
| Magnesium | 0.78 | 2.09 | 1.84 | 0.56 | 14.9 |
| Calcium | 1.06 | 3.63 | 1.94 | 1.48 | 1.12 |
| Sodium | 0.98 | 2.83 | 2.64 | 1.60 | .77 |
| Potassium | 1.33 | 2.59 | 1.42 | 2.14 | .08 |

(Clark and Washington, Goldschmidt, and Brown and Patterson)

19

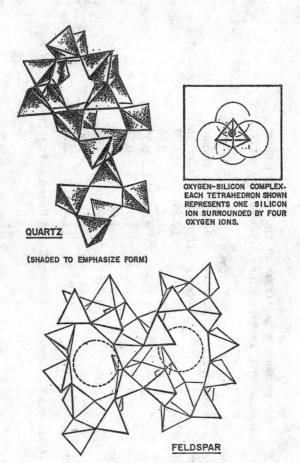

QUARTZ

(SHADED TO EMPHASIZE FORM)

OXYGEN–SILICON COMPLEX.
EACH TETRAHEDRON SHOWN
REPRESENTS ONE SILICON
ION SURROUNDED BY FOUR
OXYGEN IONS.

FELDSPAR

*Fig. 2. Oxygen is a component of most rocks and therefore, although commonly thought of as a gas, is one of the principal parts of the earth's crust. The relative proportions of oxygen to the other important elements can be visualized in these schematic draw-*

20

Rocks are *oxygen compounds*. They are composed of such different minerals as quartz, feldspar, and mica, but all these can be visualized as three-dimensional networks of oxygen held together by other atoms. When you consider the relative volumes of the crust elements, it is striking that the hard earth you walk on is for the most part an oxygen platform. This is shown on page 19 in the table of the elements that compose 99 per cent of the crust.

Scale models of the atomic arrangement in the common rock-forming materials show more vividly the predominance of oxygen. In Fig. 2 are models of the crystal structure of quartz and feldspar. The large white spheres are oxygen atoms, the small black ones silicon. Atoms of other elements bind the combinations of oxygen and silicon together in ways that form loosely bonded chains, sheets, or three-dimensional networks. These different forms of crystal structure (Fig. 2) determine how minerals break—in long splinters (hornblend) or in sheets (mica) or in shell-like shapes (quartz). Because different groupings of oxygen with silicon are basic in the structure of most common rock minerals, they are called *silicate minerals*.

---

*ings of quartz and feldspar crystals. Silicon is the next most important element, and most rock-forming minerals are mainly silicon-oxygen compounds known as silicates. (Figure taken from model structures of Dr. Tibor Zoltai, M.I.T.)*

## How We Have
### Learned about the Earth's Interior

Since ancient times earthquakes, volcanoes, and what we now recognize as evidence of evolution have fascinated thinking men. Xenophanes of Colophon (c. 500 B.C.) found sea shells in the mountains. The great Greek philosopher Aristotle (384–22 B.C.) thought winds inside the earth caused earthquakes, and he believed that rain falling on the heated earth "generated" the winds. In his *Metamorphoses* the Roman poet Ovid (43 B.C.–A.D. 17) described many geologic processes. The Renaissance's universal genius, Leonardo da Vinci (1452–1519), insisted that fossils discovered in the hills of northern Italy once had been living creatures. René Descartes (1596–1650), the French mathematician who invented analytic geometry, advanced the hypothesis that an incandescent mass like the sun cooled to form the earth with its crust enclosing a still hot nucleus.

All this was merely speculation. It was not until around 1776 (a convenient date for Americans to remember) that geology became a science. Independently, and more or less simultaneously, A. G. Werner (1750–1817), E. F. von Schlotheim (1764–1832), and William "Strata" Smith (1769–1839) discovered that rock layers can be identified because certain fossils are found in them and not in layers above or below them. This secret revealed, geologists went to work all over the world, and in a little more than a century and a half their

22

successors have built up a great body of knowledge of the physical and chemical nature and the history of the earth's crust. But they cannot investigate the mantle or core of the earth directly and still must infer the composition and physical nature of the interior regions from evidences available at the surface.

An important approach to knowledge of the earth's interior is study of the way shock waves travel through it. This branch of science is known as *seismology* and its main instrument is the seismograph, first developed in Japan in the late nineteenth century by an Englishman, John Milne. Reduced to its essentials, the seismograph is a rigid frame anchored to bedrock; a heavy weight is delicately suspended from the frame. Movements of the frame in relation to the suspended weight, which remains stationary, are recorded with electrical devices. It is sensitive enough to detect minute vibrations of the earth.

Seismological studies of the earth depend largely on earthquakes for sources of seismic waves, because dynamite explosions are not powerful enough. (Hydrogen bombs would work, but they, needless to say, are not generally available.) However, many earthquakes occur, and they have afforded much information. Seismological stations have been established in many countries to record motions of the earth's surface; a single earthquake is recorded simultaneously all around the world. Synchronization of timing devices by radio signal makes it possible to measure precisely the time of arrival of earthquake-generated seismic waves at all the stations.

Two kinds of seismic waves travel from the *focus* of an earthquake: primary (*P*) waves, which

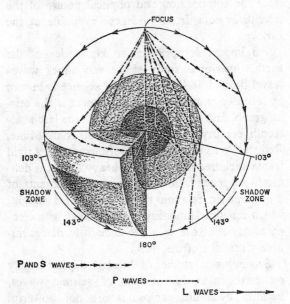

*Fig. 3. An earthquake caused by a rupture near the surface of the earth starts the types of waves illustrated in this diagram. Seismic waves travel outward from the earthquake focus as expanding nearly-spherical wave fronts. These are shown diagrammatically in the sketch as "rays," or lines drawn perpendicular to the wave fronts. These are bent by refraction when they encounter changes in physical properties at depth. The core boundary is inferred from the "shadow zone" caused by a sharp change in refraction at this depth.*

are *compression-expansion* waves like sound waves, and secondary (S) waves, which are *shear* waves transverse in motion as are the waves that travel along a stretched string. In addition there are surface waves. The P and S waves radiate in all directions from the focus, Fig. 3.

Besides indicating the existence of the Wiechert-Gutenberg discontinuity some 1800 miles below the surface, the behavior of P and S waves gives us data from which to infer something about the properties of core and mantle. It has been found that the waves are refracted (bent upward) in the region of the mantle and reflected downward at the crust. P waves propagate through both core and mantle. S waves propagate through the mantle but not through the core. From the study of shock waves in liquids and solids, it is known that liquids do not transmit shear or transverse waves like the S waves; hence, it is inferred that the core of the earth, or part of the core, is liquid.

Study of shock waves in solids has shown that the velocity of compressional waves (if you stretch a spring and let it go, the pulse traveling up it is a compressional wave) depends upon the density of the solid and its elastic constants, which are numbers that help to describe its reaction to compressing, twisting, and shearing forces. The upward refraction of compressional waves in the mantle indicates the rate at which the combination of density and elastic constants changes with depth.

In the first forty years of this century "travel time" tables for seismic waves were compiled and laboriously corrected. With them geophysicists can compute wave velocities in the earth's interior regions. Changes in wave velocity provide significant information about structure. (A change of $P$ wave velocity from about 7 kilometers a second in the crust to about 8 kilometers a second in the mantle indicated the existence of the Mohorovicic discontinuity at the base of the crust.) By studying velocity changes in relation to the densities and elastic constants of solids and co-ordinating this information with what else is known about the earth and meteorites, scientists have identified seven distinct shells in the earth and come to the conclusion that the core is composed largely of iron and the mantle largely of a rocklike material.

How do meteorites come into the picture? They are small objects from within the solar system that sometimes fall to earth. If these bodies were formed in a manner similar to the formation of the earth, it is possible that they were composed of similar material. Their composition, therefore, may give us a clue to the nature of the interior of our own planet. Two principal classes of meteorites occur: the "iron" meteorites and the "stony" meteorites. The iron meteorites are largely native iron—that is, iron not allied with other elements—and have nickel and other metallic components in minor amounts. The stony meteorites are like dark green volcanic rocks but are richer in magnesium. (See the table on page 19.)

The study of meteorites, then, leads to the assumption that the two main types of material in the earth are metallic iron and silicates (or oxides), separated as they have been in the meteorites, and it is possible to relate this hypothesis to the physical characteristics of the earth—its weight, moment of inertia, and distribution of density. Briefly, the reasoning goes like this.

The gravitational field at the surface of the earth can be measured, and with this figure the total weight of the earth is computed as $5.98 \times 10^{27}$ grams. From a measurement of the size of the earth, its average density then is found to be 5.515 grams per cubic centimeter. Now, it long has been known that the density of rocks that have risen to the surface in the form of volcanic lava and intrusive masses does not exceed about 3.3 grams per cubic centimeter. Therefore, to make up the average, density must increase toward the center. This would be expected as a consequence of pressure alone, because pressure would tend to compress the material at great depth. But theoretical considerations of the rate of change of density with pressure in materials of the stony meteorites' composition call for a different composition. If we are to account for the necessary density of the central part, a heavier material is needed. This falls in line with the hypothesis that the central part is iron.

When we take into consideration the earth's shape, a theoretical approach to the problem of density distribution is possible. You will remember that because the earth rotates on its axis the radius at the equator is greater than at the poles. Now, if we assume that the earth has responded as a perfect liquid to the combination of centrifugal and gravitational forces, and if we assume different density distributions within the mass as a whole, we can calculate the difference between the radii at equator and poles. If most of the heavier materials were concentrated in the earth's outer shells, the difference in radii would be greater than it is with most of the heavier materials concentrated at the center. Therefore, the actual measured difference of the radii at equator and poles does provide a limit to the nature of the density distribution, but it does not specify exactly what that distribution will be within the limit.

Fig. 4 summarizes the present ideas regarding density distribution in the earth, as calculated by the New Zealand mathematician and seismologist K. E. Bullen. The density distribution, coupled with seismic wave velocity data and the relative abundance of the chemical elements found in meteorites, is the most important factor in our estimates of the earth's interior. As we have seen, study of wave velocity at different depths in the mantle has given information about the existence of the discontinuities and about the relationship between elasticity and density at different depths. In later chapters we shall study the part radioactivity has played in the composition of the upper mantle region rich in such elements as silicon, aluminum, sodium, potassium, uranium, thorium, and others, and in the composition of the crust, the oceans, and the atmosphere.

### The Earth's Crust

The crust has two principal topographical features: the *continents* and the *ocean basins*. The

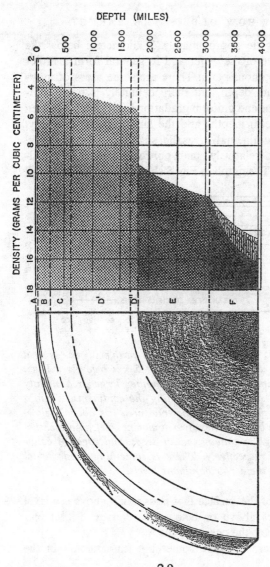

Fig. 4. Density of the earth's interior probably varies in the manner indicated by this graph, which has been developed largely from measured changes in seismic velocity with depth and a knowledge of the effect rotation has on the earth's shape.

29

basins lie approximately 5 kilometers below the surface of the water and in general form a rather monotonous plain. There are some areas of intermediate depth, but there definitely is not a continuous and uniform gradation in elevation between the ocean depths and the land surface. Similarly, the continental areas lie slightly above sea level with a relatively small portion rising to high altitude. The graph in Fig. 5 shows the strikingly small

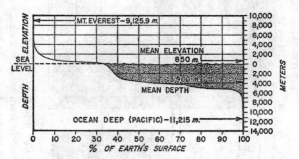

*Fig. 5. The major part of the earth's surface is at a fairly uniform level of about 5 km below sea level. Superimposed on this is another large area slightly above sea level, representing the continents. Then, a small proportion of the total area lies above, below, and in between these two dominant levels, representing the high mountains, ocean deeps, and edges of continents, respectively. There is something fundamental about these two dominant levels.*

area of the surface that rises much above sea level or lies above or below the average 5-kilometer depth of the ocean basins.

There must be something fundamental in the

30

dynamic history of the earth's surface that has caused continents to exist. Since the ocean area is four times the continental area and since the continents, as we shall see, have undergone considerable change, geologists believe that the ocean basins represent the more or less original, or primordial, surface of the earth. The continents are something different that must be explained.

The most commonly accepted hypothesis to explain the stability of the continents and the ocean basins on the earth's surface had its beginnings in the observations and conclusions of three American geologists—James Hall (1811–98), John Wesley Powell (1834–1902), and Clarence Dutton (1841–1912)—and has since been fortified by gravitational measurements all over the world.

Hall, who published a monumental thirteen-volume work on paleontology, the study of past geologic periods by means of their fossil remains, was the first to theorize that the enormous weight of tremendous accumulation of sediments along the shores of ocean basins caused depressions in the crust of the earth and that these depressions always preceded the process of mountain making. Mountains, he said, were not a product of up-heavals or convulsions of the earth but were part of continental movements.

Powell led the first boating expedition through the Grand Canyon of the Colorado River and made a study of the action of water on the rocks of the Colorado Plateau geologic district. He concluded that the process of *downwearing* was irresistible, that in time rain and rivers would wear any surface down to a *base level,* which would be the level of the ocean surface, but no farther. The higher the district, he concluded, the faster the wearing down.

Dutton, a United States Army officer, was a protégé of Powell and, like him, a student of the geology of the Grand Canyon district. He reasoned that the earth would have a perfectly smooth surface if it were composed of homogeneous material, but since the surface is not smooth, some areas of the crust must be made of lighter materials than others. In the less dense areas—mountains and plateaus—the surface would tend to bulge; in denser areas the surface would be depressed in basins which might become filled with sediments. It was plain to be seen in the mountain ranges of the West, he said, that the vast platforms on which the mountains rest had continued to rise as fast as erosion degraded the mountains themselves.

Dutton summarized this state of balance as a crustal equilibrium resulting from the force of gravity, and he coined the name *isostasy* (from Greek words meaning "standing still") for it. Gravity measurements in the Himalayas and along the Atlantic and Pacific coastlines were in accord with his theory. Plumb lines are deflected from the vertical toward the denser masses of the crust. As illustrated in Fig. 6, the surface topography of the earth results from the "floating" of materials of lower density and varying thickness in a material of greater density beneath.

Although the material of the mantle is believed to be solid, the stresses under the weight of con-

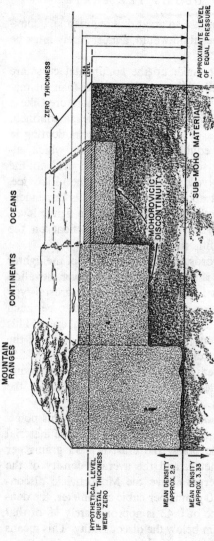

Fig. 6. Isostasy is a well-established concept which holds that the continental masses are higher than the ocean basins because they are lighter. The solid rock beneath yields slowly to the great stresses involved until heights of land are buoyed up above the general level of the earth's surface, with the height dependent upon the thickness or relative density of the lighter material.

33

tinental masses are so great that the mantle, behaving as a viscous fluid would, almost surely must be deformed.

Rocks that make up the continental crust are almost certainly lighter, or less dense, than the material of the mantle underneath. Therefore, like a great layer of shelf ice floating in a sea, a continent is visualized as a layer of lighter rocks floating in gravitational equilibrium in a medium which, although it is solid rock itself, will yield to the stresses involved, given enough time. Like ice, whose height above the water level is dependent upon the depth of the ice below the water level, so apparently the crustal materials float on the universal mantle material beneath.

In other words, when the thickness of the lighter crustal layer is small, it floats at a lower elevation than when it is large. Therefore, the crustal-type rocks in the ocean basins are believed to be much thinner than the layer in the continental areas. This is borne out by measurements of the depth of the Mohorovicic discontinuity, which show a 35-kilometer average depth under the continents and a 10-kilometer depth below sea level under the ocean basins.

A little arithmetic will help to clarify this point. For example, if the average density of the material of the mantle below the crust is 3.33 grams per cubic centimeter, and the average density of the continental crust above the Mohorovicic discontinuity is 2.90 grams per cubic centimeter, the density difference of 0.43 is approximately ⅛ of that of the medium below the discontinuity. This means that a displaced volume will have a buoyancy that will carry approximately ⅛ of its volume as additional volume above the surface of the denser medium. In other words, if the continent is floating at an elevation of 4.5 kilometers above a hypothetical mantle surface, this elevation should be supported by a submerged part that is $4.5 \times 7 = 31.5$ kilometers below this surface or about 36 kilometers below sea level. This is approximately what is observed if the Mohorovicic discontinuity is in truth the base of this submerged part.

Similarly, in the column beneath the oceans the "crust" would consist of, say, 5 kilometers of water of density 1 and 5 kilometers of rock of density 2.8, or an average density of approximately 1.9, so that the density difference in this case would be 1.43. This is more than 3 times the density difference found for the column of continental crust; so the thickness would be expected to be ⅓ as much. This conclusion is again supported by the measured depth of the Mohorovicic discontinuity under the oceans—namely, 10 kilometers. If these features of topography were truly floating in gravitational equilibrium in this manner, the measurement of gravity over the surface of the earth (corrected for altitude and latitude) should all be the same. This is found to be in large part true, except for zones of disturbance in which active mountain-building processes have gone on in the recent past so that isostatic equilibrium no longer exists.

Commonly located along the margins of continents, but not always, these zones, which are out of equilibrium and distinctly different from the stable oceanic and continental areas, are of great interest in geology because they tend to throw light on the possible origin and development of mountains and continents in the past history of the earth.

## Mountains and Volcanoes

We can see that there is a need for a large expenditure of energy in the formation of these belts of mountains. If some dynamic process develops a region that is out of gravitational equilibrium, there will be continuing stresses and yielding in the direction of restoring the equilibrium. Consider again the example of the mass of ice floating on water: If the ice above the water level is melted by warm winds and sun, more ice will be fed upward by the entire mass rising as the load is reduced, until finally the ice near the bottom of the mass, originally deep under water, will be exposed at the surface.

At least some mountain belts, like the Sierra Nevada range in California, appear to owe their elevation to deep roots of lighter material. As the mountains are eroded by rain and ice and chemical decomposition, the root would be expected to be out of gravitational equilibrium and rise like the ice mass. But the mountains can be worn down only to sea level, which is several kilometers higher in elevation than the ocean floor; so the root would stop rising, as we have seen, at a depth of about 35 kilometers. Because of the presence of the ocean, therefore, it is possible to have permanently stable masses like continents on the earth's surface, and the explanation of the Mohorovicic discontinuity becomes evident.

Thus we see that whenever a mass of lighter material develops, it will be unstable until worn down to sea level and until it reaches gravitational balance. After that time it should be a permanent fixture on the earth's surface, unless processes are at work other than those considered here. This fact is borne out by the observation that although large areas of the continents are very old, they have not disappeared or sunk appreciably but have remained almost at sea level since they were first eroded to this level and the base of their roots had risen to the 35-kilometer mark.

How are these lighter masses formed? This question is strongly debated among geologists, but one hypothesis is that when a part of the upper mantle melts, as a result of insufficient removal of the heat from radioactivity, a separation of the elements occurs. The elements that form stable minerals at higher temperatures and pressures remain at depth, while the elements that become semi-liquid at these temperatures and pressures rise toward the surface. Not only is there some separation of these semi-liquid components toward the surface, but they also recrystallize near the surface into mineral forms which are less dense than those containing the elements originally at depth before the melting happened.

For example, sodium, calcium, aluminum, and silicon combine with oxygen to form the relatively light feldspar structure in the crustal zone near the surface, but undoubtedly they were held in more dense crystal structures before they were released from their earlier mantle location.

Let us say, therefore, that one way in which the lighter continental masses can come into existence is by a heating of the upper mantle to a temperature of at least partial fusion, with a resultant upward separation of certain elements and recrystallization into less dense minerals. We have already noted how, by the build-up of thick masses of sediment at the margins of continents, it is possible for the flow of heat to be impeded in these areas so that melting at depth can occur. And we know too that the greatest mountain belts have occurred along the former margins of continental areas. There is evidence in some parts of the earth that the continents are growing in size outward, by adding stable new belts at their periphery, with the material coming in part as sediments from the erosion of the continent, but mostly from the mantle below.

Of course, not only are there many variations on these ideas, but also many different hypotheses; and it is probable that several different processes have been active in different areas and at different times in earth history. Nevertheless, all possible processes have the common requirement of energy. So let us examine the sources of energy and their limits in more detail.

# CHAPTER II

# Radioactivity

At this instant the reader, if he weighs about 170 pounds, is being bombarded internally by approximately 700,000 damaging bullets per minute. These bullets are sufficiently powerful to break chemical bonds, modify molecular structures, and even destroy cells. The reader is not conscious of this constant disruptive bombardment because his body has the power to repair the damage much more rapidly than it is caused. But though the truth is not completely understood, it is possible that these natural radiations may be a potent factor in the life span of a living organism or in the evolutionary process whereby organisms change slowly throughout time into different forms.

To measure radiation dosage, science commonly uses a unit called the *rad,* which represents the quantity of energy dissipated per gram of matter, or:

$$1 \text{ rad} = 100 \text{ ergs} = 2.4 \times 10^{-6} \text{ calorie per gram.}$$

(A unit called the *rem* also is used; it represents an amount of energy absorbed, with correction for the relative biological effects of the different kinds of radiation.)

If we are to gain a perspective on the relative importance of the possibly injurious sources of radiation—the first that springs to mind is, of course, fallout from tests of nuclear devices—we shall find it interesting to compare the doses that human bones receive from *all* sources. The table opposite, summarized in 1958 by R. A. Dudley, gives instructive estimates, which no doubt will be revised upward as our studies continue.

From this table you can see that the doses received from natural radionuclides of the uranium and thorium series and potassium, both in the body and the outside environment, and from cosmic rays, which always have bombarded inhabitants of the earth, have been roughly equaled by the new sources of radiation, such as medical X-rays and fallout. Furthermore, commercial power generation from fission is expected to reach 30,000 megawatts by 1975, and it will produce fission products at a rate equal to 300 megatons of bombs per year. This greatly exceeds weapon-testing rates; it will be important to prevent even a small proportion of these fission products from escaping.

(The isotopes in the first line of the table are called *radionuclides* because they are radioactive and because they are species of atoms characterized by the constitution of their *nuclei;* that is, by the numbers of protons and neutrons their nuclei contain. The symbol K stands for potassium, Ra for radium, and Pb for lead.)

ESTIMATED SKELETAL DOSE RATES OF MAN

| Source of Radiation | Avg. Skeletal Dose Rate millirad/year | Comment |
| --- | --- | --- |
| Naturally radioactive isotopes present in the bone ($K^{40}$, $Ra^{226}$, $Ra^{228}$, $Pb^{210}$) | 34 | Average value only. Variations within a factor of 3 to 5. |
| Strontium 90 and Caesium 137 from fallout | 3 (1957, or 12 for 1975) | From past explosions only. |
| Cosmic rays | 30 | Value near sea level at higher latitudes. |
| Naturally radioactive isotopes present in environment | 45 | Typical value for sedimentary rock area; allowing for building materials. |
| Gamma rays from fallout | 0.5 | In U.S., 1951–56 |
| Medical X-rays | 75 | Value very variable and inaccurately known. |
| Total | 200 | |

To safeguard individuals working with radioactive materials, the Atomic Energy Commission has proposed regulations to limit the basic permissible tissue dose for exposure to any ionizing radiation. For an indefinite period of years the dosage for internal organs could be 100–300 millirads per week. Thus, the natural and man-made radiations still amount to only $\frac{1}{50}$ of the AEC's specified limits for the safety of workers.

### Radioactive Fallout

Localized concentrations of short-lived radioactive isotopes formed in bomb tests received considerable publicity a few years ago. For instance, after the H-bomb tests in the Bikini Atolls in March 1954 considerable amounts of radioactive zinc 65 and other isotopes were found in tuna fish arriving in Japanese ports. Since the Japanese people derive 90 per cent of their protein diet from sea products, the hazard was considered so serious that the products were not easily sold on the market. In the month following the bomb tests the market prices in the principal fishing port of Misaki dropped to less than one half. Contaminated fish were caught at great distances from Bikini, even in waters surrounding Japan.

The early scares over such fallout instances now have given way to careful estimates of the long-range effects of nuclear fallout and contaminations, and the estimates are based on many measurements and sober calculations. Most of the present-day concern is with the level of concentration of a few long-lived isotopes, of which much the most important is strontium 90. Since this isotope, which has a 28-year half-life, is gradually brought to earth by rain over a period of several years following nuclear explosions, its concentration will gradually increase. Cows eating grass take it up from the ground, and it eventually gets into the human system through the consumption of milk and dairy products. To a small degree it supplants calcium in the bones.

The 1958 report of the Advisory Committee on Biology and Medicine, of the National Academy of Sciences, came to the conclusion that the population of the United States should not, as a genetic safeguard, receive a radiation dose exceeding an upper limit of 10r in thirty years. At this writing (1959), United States residents, it is estimated, have on the average been receiving from fallout over the past five years a dose which, if weapons testing were continued at the same rate, is estimated to produce a total thirty-year dose of about 0.1r, or $\frac{1}{100}$ of the estimated permissible limit. The average world-wide thirty-year dose was estimated to be considerably lower. Thus, the report suggested that there was not yet any cause for alarm, but in view of the adverse repercussions caused by the testing of nuclear weapons, it recommended that tests be held to a minimum consistent with scientific and military requirements.

In short, therefore, radioactivity is all around us and we cannot escape it. The level of background radioactivity has been part of the environment of living creatures, and the evolution of life on the earth has become thoroughly adjusted to it. Man-made concentrations or disturbances in this realm could possibly cause harmful concentrations, but with the correct scientific vigilance it should be possible to keep control of this danger, unless the matter becomes one of malign intent in war. Let us, therefore, not fear radioactivity, but use it to our advantage.

## Radiation and Particles

What is the nature of these common radiations and fast-moving particles? Where do they come from? How does the physicist produce them and use them? Before we go on to examine, in subsequent chapters, radioactivity's role in measuring the age of the earth and the part it has played in shaping our destiny, we should have at least a general understanding of some fundamentals.

First we must have a glossary of terms. In the strange new world of nuclear physics the scientist finds himself able to observe phenomena and measure quantities and effects that he does not understand fully. Names are given to these observed entities while they are being studied, but generally these names do not last very long because the entity is found to be made up of more basic entities; the name becomes superfluous. It is highly instructive for the student of physics to study the historical development of every phase of the science. We shall do this briefly in later pages, but first let us have a list of the present names given to the quantities in our subject and omit the confusing earlier terminology. For example, terms such as canal rays, cathode rays, and positive rays have historical interest to the physicist, but actually are obsolete.

Therefore, keeping in mind that giving something a name merely isolates one or a few of its properties and does not describe it fully, we shall list some commonly used terms and their principal characteristics.

The *nucleus* of an atom is currently described, in terms of two particles, *neutrons* and *protons,* which exist in roughly similar proportions in most of the elements. As the name implies, the neutron is an electrically neutral particle, while the proton has a positive electric charge of $4.8029 \times 10^{-10}$ electrostatic units. Because energy is required to develop an electric charge, the bringing together of a number of protons will increase the total charge density, or energy, in a nucleus. As a result, the nucleus would tend to fly apart if it were not for some binding force, little understood, that keeps the particles together. The particles are held so close by these short-range binding forces that the density of the nucleus is about $2 \times 10^{14}$ grams per cubic centimeter.

The difference between an atom of one element and the atom of another element is the difference in total positive charge of the nucleus, or the number of protons in it. The total number of neutrons plus protons in a nucleus is referred to as the *mass number* A. The total number of protons or positive charges in the nucleus is known as the *atomic number* Z. For example, uranium 238, which is the largest naturally occurring nucleus, contains 92 protons and 146 neutrons. A common notation that defines every individual atomic species has the atomic number as a subscript before the symbol for the element and the mass number as a superscript following it; for instance, $_{92}U^{238}$ means that this atomic nucleus has 92 protons and $238 - 92 = 146$ neutrons.

Atoms that have the same number of protons but different numbers of neutrons are still almost chemically identical and are called by the same element name. They are referred to as *isotopes* of the element. For example, oxygen has three naturally occurring isotopes, $_8O^{16}$, $_8O^{17}$ and $_8O^{18}$.

Like the cocking mechanism on a spring-loaded gun, the binding forces that hold the protons and neutrons together in a nucleus must act only at very short range. Thus, when a nucleus is split by some means, the fragments fly apart with great velocity. When too many protons are packed together with neutrons and the binding forces are close to their limit of being able to hold the assemblage together, the normal vibrations within the structure may occasionally exceed the limit of a bond, and a part of the nucleus flies off spontaneously. This is referred to as *radioactivity,* or the radioactive *breakdown* of the nucleus.

In the simple loss of a fragment of the nucleus one of the most usual particles emitted is always made up of two neutrons and two protons. This is the same as the nucleus of the helium atom and is known as an *alpha particle*. The loss of an alpha particle usually leaves the nucleus in an *excited state,* and it does not settle down to a stable *ground state* until it has further emitted one or more *gamma rays.* A gamma ray, like a single photon of light, has properties that make it behave like electromagnetic radiation in some circumstances and a small bundle of energy (equivalent to a particle of matter) in other circumstances.

Another way in which an unstable nucleus can acquire greater stability is by a *transformation* in which the nucleus changes its charge by one and emits an *electron,* together with a *neutrino.* The high-speed electron emitted radioactively from a nucleus is known as a *beta particle.*

In an alpha transformation the mass of the alpha particle plus the mass of the remaining nucleus does not quite equal the mass of the original nucleus. The difference in total mass before and after the *event* represents the mass-equivalent of the energy expended in the alpha particle and gamma rays. The energy and the mass differences are related by the Einstein equation $E = mc^2$. The energy $E$ is given as the kinetic energy of the alpha particle and recoiling parent nucleus and as the radiation energy of the gamma rays; $c$ is the velocity of light, equal to about $3 \times 10^{10}$ centimeters per second; $m$ is the difference between the mass of the nucleus before the alpha particle was emitted and the mass of the nucleus plus alpha particle afterward. In the beta transformation the energy of the beta particle plus concomitant measurable radiation is not sufficient in terms of mass equivalent of energy to make up the difference in mass before and after the transformation. As a result, the neutrino has been postulated by the Swiss physicist Wolfgang Pauli to make up the difference. The existence of the neutrino has been hard to prove, but in recent experiments it has been detected.

Finally, a nucleus may occasionally undergo another kind of transformation in which it captures one of its associated electrons and emits a neutrino, changing the atom's charge by one. This is referred to as *K-electron capture*. This transition is important in geology, because it provides the basis for the argon-potassium method of measuring geologic time, which we shall describe later in the book.

To summarize, we see that there are alpha particles, beta particles, and gamma rays involved in the radioactivity about us. The alpha particles are heavy, being made up of two neutrons and two protons, and although very energetic, travel only about ten centimeters in air, or the width of a pin point in solid matter. The beta particles are electrons, which travel a little farther than alpha particles in solid matter, but are quickly stopped by their interaction with the surroundings. The gamma rays are like unusually powerful X-rays and will penetrate matter for a distance of several centimeters, but are largely absorbed in the first few centimeters traversed.

With these few brief terms in mind let us now return to the interaction of these radiations and particles with matter.

## Absorption of Gamma Rays

Almost everyone has become familiar with the fact that gamma radiations can be detected with portable devices. The prospector roams the hills with a portable Geiger counter or scintillation counter; workers in any area with radiation hazard wear film badges to indicate their exposure to radiation; civil-defense organizations instruct radiation monitoring teams in the use of portable detection instruments.

The range of gamma rays from a radioactive source material depends upon the amount of matter traversed. It is not necessary to have a lead shield about a highly radioactive source, if the source can be kept where people do not get close to it. In other words, radiation intensity is reduced by distance, as well as by shielding materials.

Gamma rays are absorbed by matter. By matter we mean atoms of different kinds in gaseous, liquid, or solid state. The volume of an atom is made up almost entirely of empty space traversed by electrons. The interaction of gamma rays with matter is almost entirely their interaction with electrons. The mass of the electrons is negligible compared to that of the very dense nuclei, but these nuclei make up only a very small part of the field of view. As the total number of electrons usually is equal to the total number of protons in the nuclei, and the number of these is almost proportional to the total mass of the nuclei, the density of electrons in any material is almost proportional to the mass of that material measured by its weight per cubic centimeter.

Thus, lead having a little more than ten times the density of water, gamma rays will be absorbed more than ten times more effectively by lead than by water.

There are two principal ways in which gamma rays interact with electrons.

First, a gamma ray may strike an electron and bounce off it at an angle, imparting a velocity to the electron in the other direction. The kinetic energy given to the electron is taken out of the gamma ray, so that the ray continues on its path with reduced energy. The term "bounce" is a strange one to use for something with the properties of electromagnetic radiation, but when you remember that the electron is a charged entity which acts as either a particle or a wave, depending upon the way it is observed, and that the gamma ray is a coupled electric and magnetic field occupying an extremely small and impossible-to-imagine cross-sectional area in its trajectory through space, the phenomenon cannot be anything but strange in the first place. Let it be said simply that the gamma ray loses some of its energy, changes its direction, and knocks an electron out of its atomic orbit with a velocity that will damage surrounding electronic structures. This effect is known as *Compton scattering*. It is illustrated schematically in Fig. 7. The American physicist A. H. Compton, working with X-rays, discovered it in 1923.

Second, there is an interaction known as the

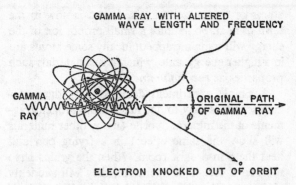

GAMMA RAY WITH ALTERED WAVE LENGTH AND FREQUENCY

GAMMA RAY

ORIGINAL PATH OF GAMMA RAY

ELECTRON KNOCKED OUT OF ORBIT

*Fig. 7. Gamma rays can dislodge the outer electrons of atoms, losing some of their energy and giving enough to those electrons for them to affect other atoms. This schematic representation of Compton scattering is not a true picture of the atom and gamma rays but does help us to discuss the phenomenon.*

"photoelectric effect." This effect happens with photons of light as well as photons of gamma radiation. (A photon is a *quantum,* or discrete "packet," of radiation.) When a gamma ray has lost most of its energy through Compton scattering, it has a high probability of losing the rest of its energy in complete absorption in an atomic electron. Then the gamma ray simply disappears and its entire energy is imparted to the electron.

The sum of these two effects is that gamma rays lose their energy until they disappear, the energy being given up to one or many electrons, which in turn lose their energy by striking or interacting with other electrons until, finally, the energy has been changed into a general excitation of the atoms of the matter traversed, and is now in the form of heat. Sometimes a small proportion of the energy will remain trapped; locally some atoms are in a higher energy state, which gives materials such properties as *thermoluminescence*.

A simple experiment demonstrating thermoluminescence is easy to do. Place some broken fragments of the mineral fluorite (many other minerals will show the same effect) in a frying pan and heat them in a dark room. When the grains have reached a certain temperature, they will suddenly glow for an instant with a dull light. This light is given off as electrons fall back into positions from which they have been driven by gamma radiation over a long period of time. The gamma radiation comes from radioactive impurities in the fluorite.

We see, therefore, that the attenuation of gamma rays is dependent upon two laws. First, like light, gamma rays decrease in intensity with the square of the distance from a point source. Purely a geometrical attenuation, this can be calculated for any shape of source. We consider the source as a large number of small point sources and add the effects of each.

Secondly, the attenuation of gamma rays is one of probability of striking electrons. In any specific case the probability is proportional to the number of electrons present in a unit volume and also to the number of gamma rays present. Let us take, for example, a parallel beam of gamma rays traversing a homogeneous material in which the electron density is essentially constant. This is shown diagrammatically in Fig. 8.

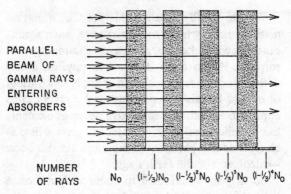

*Fig. 8. The intensity of a beam of gamma rays penetrating a series of equal plates of absorbing material decreases* exponentially, *as the number of plates (or total thickness) is increased, as shown above.*

Let us say further that the number of interactions of gamma rays with electrons in the first centimeter of travel through the material reduces the number of gamma rays by a certain factor—say, by one third. In the next centimeter of travel we start with only two thirds the original amount of gamma rays; therefore, the probability of interactions in the second centimeter of travel is one third of two thirds, or two ninths, leaving only four ninths of the original gamma rays to continue. In other words, in every successive centimeter of travel the number of gamma rays emerging from that centimeter of material will be a constant proportion of the number of gamma rays entering the material. This proportion will be set by the electron density of the material.

We can readily describe this process in mathematical terms, which, as we shall see, have significance beyond the description of gamma-ray absorption. As so often happens in physics, a law derived for one relationship may apply to a variety of others. In this instance our final equation also applies to the rate of radioactive decay of elements like uranium and thus, as later chapters will disclose, is doubly pertinent to the subject of our book —calculation of the earth's age.

So, if you would care to come with me on a brief side excursion to examine what is commonly known as the "exponential law," I think you will find it fun and pleasantly simple, but if not, you can join us a bit further on, and we shall still arrive at our main destination together.

First we shall call the total distance the gamma rays travel "$x$" and say that there are "$n$" number of equal short distances $\triangle x$ in $x$, just as there are 10 one-inch intervals in 10 inches, 10 being the $n$, one inch the $\triangle x$, and 10 inches the $x$ ($\triangle$ is the Greek letter *delta*). We shall use $N_0$ (called "$N$ subscript $0$") for the number of gamma rays entering the material, and $N$ for the number of rays left at any point in the material. We shall use another Greek letter $\mu$ (*mu*) for the reduction factor, which in the foregoing example we arbitrarily set at one third. Now we are ready to start.

In the short distance $\triangle x$ the material will absorb $\mu \triangle x$ times the number $N_0$ rays that entered, or $N_0 \mu \triangle x$. The number of rays left will be the original total minus those absorbed, or

$N_0 - N_0\mu\triangle x$, which can be expressed as $N_0$ times $1 - \mu\triangle x$, or $N_0(1 - \mu\triangle x)$.

In traversing the second short distance $\triangle x$ the rays again will be absorbed in the same constant proportion. So, we multiply $N_0 - N_0\mu\triangle x$ by $\mu\triangle x$ to find the number of rays absorbed. We get $N_0\mu\triangle x - N_0\mu^2\triangle x^2$. Subtracting this from the number of rays that *entered* the second short interval, we find the number left at the end of the second interval:

$$N_0 - N_0\mu\triangle x - (N_0\mu\triangle x - N_0\mu^2\triangle x^2)$$
$$= N_0 - N_0\mu\triangle x - N_0\mu\triangle x + N_0\mu^2\triangle x^2$$
$$= N_0 - 2N_0\mu\triangle x + N_0\mu^2\triangle x^2$$
$$= N_0(1 - 2N_0\mu\triangle x + \mu^2\triangle x^2)$$

which is $N_0(1 - \mu\triangle x)^2$, the number of rays left at the end of the second $\triangle x$ distance. We could go on through a third $\triangle x$, getting $N_0(1 - \mu\triangle x)^3$, and a fourth, $N_0(1 - \mu\triangle x)^4$, but it is better to use general terms rather than specific terms. So let us just consider $n$ distances of $\triangle x$. After $n$ intervals of $\triangle x$, the number of gamma rays left will be $N_0(1 - \mu\triangle x)^n$.

Now, as we said earlier, there are $n\triangle x$ distances in the total distance $x$; so:

$$n\triangle x = x \text{ or } \triangle x = \frac{x}{n}$$

To tidy matters up a bit more, let us substitute $\frac{x}{n}$ for $\triangle x$ in our general expression of the number of remaining gamma rays. Thus, $N_0(1 - \mu\triangle x)^n$ becomes $N_0(1 - \mu\frac{x}{n})^n$, and that brings us to the equation we were seeking, the law stating the number of remaining gamma rays $N$ at any point in the material. We have (see Fig. 8)

$$N = N_0(1 - \mu\frac{x}{n})^n$$

To get an accurate statement of this law, we shall have to make the $\triangle x$ intervals very small. The smaller they become, the larger $n$, the number of those intervals, must become. (There are more hundredths of an inch in 10 inches than there are tenths of an inch.) The ultimate accuracy comes when we let $\triangle x$ approach zero; then $n$ necessarily approaches infinity.

We shall not prove it here, but the number 2.718, known as $e$, has the valuable mathematical property that $e$ raised to the $x$th power equals the limit of $(1 + \frac{x}{n})^n$ when $n$ approaches infinity. We write it

$$e^x = \lim_{n \longrightarrow \infty} (1 + \frac{x}{n})^n$$

If we apply this to our friend $N = N_0(1 - \mu\frac{x}{n})^n$, we get the exponential equation:

$$N = N_0 e^{-\mu x}$$

This equation, as I have said, tells us not only about gamma rays but also about the disintegration of uranium. If we make $N_0$ the original number of uranium atoms millions of years ago and $\mu$ the proportion of atoms disintegrating each second, then $N$ will be the number of atoms of uranium left at the end of any time $x$. Hence its value in calculating the age of our planet.

And now let us return to our laboratory—the earth and the particles that compose it.

## Alpha and Beta Particles, and Heat

When an alpha particle travels through matter, its two protons represent a fast-moving positive charge, which reacts with the electrons in its path, pulling them out of their orbits and imparting an additional velocity to them. The mass of the particle is so great, however, that the interaction of the electrons does not change the direction of the alpha particle to any extent, and it gradually loses energy by accelerating the electrons in the path of its travel. As before, these accelerated electrons gradually transfer their energy to a general excitation of the electronic and atomic structures. This excitation we know as heat.

There will also be some loss of energy from the interaction of the positive charge with the nuclei of the atoms of the matter traversed. But in most cases it is not until nearly at the end of its path that the alpha particle interacts with atomic nuclei. Except for the rare instance when the alpha particle strikes a nucleus in the earlier part of its travel, it usually travels until its velocity drops to a small fraction of its original velocity before it collides or interacts with atomic nuclei. At this time it has sufficient energy left to knock a nucleus out of its position in a crystal structure, and this nucleus gains enough speed from the collision to knock other nuclei out of position until all the energy has been dissipated.

The recoil imparted to the parent nucleus from which the alpha particle came is sufficient to disrupt other atomic nuclei in the material. Because the speed of the recoiling parent nucleus is low compared to the velocity of the alpha particle, it spends most of its energy disrupting other atoms instead of electrons alone. As a result, most of the damage done in the alpha event is done by the recoiling parent nucleus. Several thousand atoms are disrupted at the two ends of the alpha particle's path. Almost all the energy is lost as heat. The damage and excitation of the structure is so intense locally that it is equivalent to a temperature of tens of thousands of degrees centigrade in the disrupted volume. Depending upon the type of material, this damage will remain permanently in the structure, or the structure will gradually return to its crystalline arrangement through the passage of time. Either way very nearly all the energy goes immediately into heat.

The damage caused by a beta particle is rather similar to that of a gamma ray. It will interact with electrons along its path but not disrupt atoms, except by disturbing the bond between atoms and thereby possibly causing chemical rearrangements. It loses energy to the surrounding electronic structure by accelerating the electrons, and again all but an insignificant amount of the energy goes eventually into heating the structure.

So, we see that all radioactive disintegrations

eventually result in a supply of heat to the surrounding material. The amount of heat supplied can be calculated from the known energy of the different particles and radiation. Now, after this brief survey of radioactivity, we are ready to seek the significance of this heat in geologic history, to learn how these radioactive processes have profoundly changed the character of the earth, making it a hospitable place to live on instead of a barren, desolate waste.

# CHAPTER III

# The Radioactive Earth

Without the heat from radioactivity it is probable that we would have had no atmosphere or oceans. Even if the ocean had existed, no land would have risen above it. Indeed, it is probable that the earth would have had a bare, rocky surface like the moon's, scorched by the sun in daytime and bitter cold at night. You who read this book would never have been born.

But the story of the earth is a story of heat. Throughout earth history large amounts of energy have been continuously expended in mountain building, volcanism, and other activities which have formed the continents, oceans, and atmosphere. Except for the actions of the surface agencies, driven by heat from the sun, the energy comes from the interior of the earth and must have been at one time in the form of heat. To try to explain the occurrence of this thermal energy, we must consider two principal sources: the heat inherited from the formation of the earth and the heat generated in the breakdown of radioactive elements.

There are many arguments in favor of believing that the earth formed at a relatively low temperature. If this is true, a uniform distribution of the radioactive elements that we estimate were contained within the earth would have heated it sufficiently to have caused it to melt or partly melt. It is purely by chance that the sequence of events which we believe followed was such that the bulk of the earth stopped heating up again and remained fairly stable.

These purely chance events are as follows. First, if the mantle of the earth is solid and there is no convection (transfer of heat by movement of fluid material) in it, it must lose heat by the slow process of conduction in the upper regions. If there is convection, from melting or otherwise, the heat can be rapidly transported to the surface. Therefore, the temperature can never get very much above that necessary for melting. In the lower regions heat may be carried out by radiative transfer.

Second, as soon as melting begins, there probably would be a migration of the molten radioactive substances upward because they crytallize at lower temperatures than compounds of magnesium, silicon, and iron. They would be forced upward in the liquid as the solid material settled downward.

If in any region the heat-producing elements have not moved close enough to the surface, the temperature will rise locally to the melting point and a further upward migration will occur. Eventually they will have come close enough to the surface so that there will be no further melting.

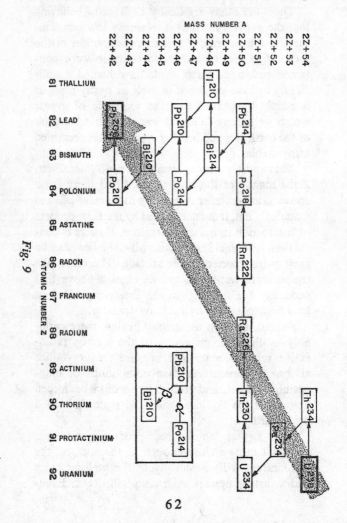

Fig. 9

62

Enough of the heat generated will be lost by conduction to the surface so that a stable solid mantle remains. Gradually, following that time, the radioactive elements will decay slowly, and their heat production will diminish. This would tend to stabilize the mantle so that it is at some temperature below the melting point of its most fusible components. But if the chemistry of the radioactive elements had been otherwise and they had settled into the core, the earth would be continuously melting, losing heat by convection and solidifying again.

## The Major Heat-Producing Elements

Let us now examine the amount of heat given off by radioactive elements and estimate what abundance of these elements would cause melting in the mantle. The element uranium breaks down through several stages to form a stable end product, lead (see Fig. 9). As it undergoes successive transformations toward this stable end product, the isotope uranium 238 gives off 8 alpha particles as well as numerous gamma rays and beta particles. Summing up the energies of all these emitted particles and rays, we find that a total of 47.4 Mev (million electron volts) of energy has been ex-

Fig. 9. (opposite) Uranium 238 decays spontaneously to form thorium 234, which in turn breaks down into protactinium 234, and so on until the procession stops at lead 206, which is stable. Some of the transformations are accompanied by alpha particle emission and some by beta particle emission.

pended for each atom of uranium 238 that breaks down to form an atom of lead 206. Since 1 Mev is equivalent to $3.83 \times 10^{-14}$ calories, it can be calculated that one gram of uranium in equilibrium with its daughter products is continuously giving off 0.71 calories per year. Similarly it can be calculated that the isotope uranium 235, of which atom bombs are made, is giving off 4.3 calories per gram per year when in equilibrium with its daughter products. Thorium and its series give off 0.20 calories per gram of thorium per year. The only other important heat-producing element is the isotope of potassium, $K^{40}$. This gives off beta particles and gamma rays at a rate that yields $27 \times 10^{-8}$ calories per gram of total potassium per year.

Average granite and volcanic rock contain approximately the following amounts of these radioactive elements:

| Rock Type | Uranium parts per million | Thorium parts per million | Potassium % |
|---|---|---|---|
| Granitic rocks | 4 | 14 | 3.5 |
| Dark-colored volcanic rocks | 0.6 | 2 | 1.0 |

Thus the radioactive components of the average granite can produce 7 microcalories of heat per gram per year. Other rocks that make up the bulk of the crust produce somewhat less heat than granites, and it is estimated that the average rock in the crust above the Mohorovicic discontinuity probably produces about 2 microcalories of heat per gram per year.

There is a measurable amount of heat continuously flowing to the surface of the earth. Measurements over continental areas have indicated that this amount averages about 1.2 microcalories per square centimeter per second. The amount of heat flowing in oceanic areas has been difficult to measure, but several measurements have been made. This is done by dropping a probe from a ship so it penetrates the mud on the bottom of the ocean for some distance. Refined temperature devices are then used to record the difference in temperature between two points on the probe. By determining the thermal conductivity of the mud, it is possible to calculate the amount of heat flowing upward from the earth into the ocean water. Surprisingly, it turns out that approximately the same amount of heat is flowing from the interior of the earth in the oceanic areas as in the continental areas; namely, about 1.2 microcalories per square centimeter per second.

From all these figures it can be calculated that the average continental crust down to a depth of 35 kilometers produces about ½ microcalories per square centimeter per second from the radioactive breakdown of uranium, thorium, and potassium. This is about one half of the observed heat flow to the surface. It means that only about one half of the heat flowing to the surface comes from a depth of greater than 35 kilometers.

If you measure the amount of heat flow and estimate the thermal conductivity of the materials in the crust and below the crust, it is possible to estimate the increase in temperature with depth.

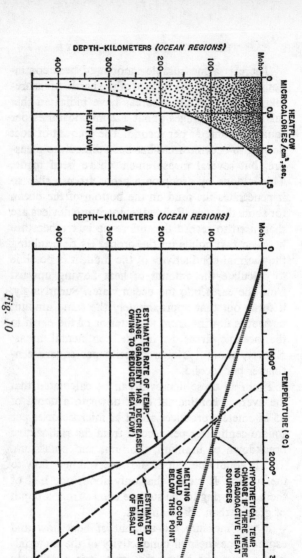

DEPTH—KILOMETERS *(OCEAN REGIONS)*

HEATFLOW MICROCALORIES/cm³, sec.

HEATFLOW

DEPTH—KILOMETERS *(OCEAN REGIONS)*

TEMPERATURE (°C)

ESTIMATED RATE OF TEMP.
CHANGE (GRADIENT HAS DECREASED
OWING TO REDUCED HEATFLOW)

MELTING
WOULD OCCUR
BELOW THIS POINT

HYPOTHETICAL TEMP.
CHANGE IF THERE WERE
NO RADIOACTIVE HEAT
SOURCES

ESTIMATED
MELTING TEMP.
OF BASALT

*Fig. 10*

66

Fig. 10 shows some estimates of the temperature-depth relationship. Note that the production of the heat in the crust greatly reduces the thermal gradient (rate of heat flow) at depth. It follows that the temperature at depth is very much less than would be expected if one simply measured the temperature in deep mines or other openings in the earth near the surface and extrapolated this information to great depth. In fact, if there were no radioactivity in the crustal rocks, the observed temperature gradients at the surface would require that the mantle be molten at a shallow depth; this is not in agreement with the known geological stability of the region. The facts, therefore, support the hypothesis that the radioactive components of the earth are largely concentrated in the near-surface layers.

By calculating the temperature at which materials would be molten at depths of 100 or 200 kilometers, it is possible to estimate how much radioactivity must be in the near-surface rocks in order to keep the temperature gradient within

---

*Fig. 10. (opposite) The temperature gradient, or maximum rate of change of temperature in a body, is proportional to the heat flow and inversely proportional to the conductivity. Near the surface of the earth, for example, the heat flow is $1.2 \times 10^{-6}$ cal/cm$^2$/sec; the conductivity is .007 cal/cm/degree C/sec. So the gradient is $1.2 \times 10^{-6}/.007 = 17 \times 10^{-5}$ degrees C per cm or 17° per km. Heat-producing elements lie between the surface and any point below; the heat flow at depth is therefore less and so is the gradient.*

known bounds. Attempts to do this have indicated that at least 0.2 part per million of uranium and 0.7 part per million of thorium on the average must be in the rocks down to 100 kilometers depth under oceanic regions. Since these amounts of radioactive elements would supply much of the observed heat flow to the surface, there must be little heat left flowing from the interior.

Thus we arrive at two important conclusions. The first is that very little of the original heat stored in the earth at the time of its formation is being lost, and, therefore, the earth is not cooling down at an appreciable rate, if at all. Secondly, we conclude that the major part of the earth's heat is coming from the breakdown of radioactive elements. Since almost all the breakdowns occur within the near-surface regions of the earth, it is reasonable to infer that some process has moved the radioactive elements to this location from a presumably homogeneous distribution at the time of the earth's origin.

## Migrations through the Mantle

Again we see the need for some process to have brought up from within the earth the materials that make up the oceans and atmosphere and the radioactive elements. In support of this requirement, we see that uranium, thorium, and potassium do in fact belong to the group of elements that form compounds of rather low stability and therefore would be most likely to move to the outer part of the mantle in any process in which partial melting was involved.

It is interesting that these conclusions do not violate the concept of an earth that is composed of materials similar to the iron and stone meteorites. The proportion of radioactive components in iron meteorites is very small indeed and would contribute a negligible amount to the heat of the earth if it were made of similar material. The amount of radioactive components in stony meteorites has been measured carefully, and it is a striking coincidence that the amount corresponds very closely with that needed to give rise to the observed heat flow in the earth if it were uniformly of stony meteoritic composition. The fact that the radioactive components have migrated to the outer part of the mantle does not alter this interesting and supporting evidence.

Thus we see an earth in which the central part is rather slowly changing, if at all, in temperature and losing heat to the outside very slowly. Near the surface, however, there is an important balance in which the heat produced by radioactive elements can flow to the surface without causing melting unless the system is disturbed in some way. If at the margin of a continent the accumulation of sediments formed a low-conductivity blanket which also contained added amounts of heat-producing elements, this might be enough to cause melting at a depth of 100 or 200 kilometers and give rise to volcanic activity and other effects related to mountain building.

There has been much discussion and difference

in opinion about the possibility of major convective overturns in the mantle down to the core boundary as a result of inhomogeneous distribution of heat sources. This process of convection could give rise to surface activity also and could be the cause of mountain-building events. In either case it is the heat from radioactivity that provides most of the energy for the dynamic events that have occurred at the earth's surface throughout geologic time.

# CHAPTER IV

# Measurement of Absolute Geologic Time

Since the first hot days when the thickening atmosphere swept in gales across the earth, alternately drenching it with rain and drying it with searing winds, its surface has been carved and molded, buried and exposed again. Unlike the moon, which has no atmosphere and has retained its primordial features until today, the earth has developed a challenging complexity of eroded surfaces and filled basins. The energy for these superficial alterations has come from the sun, and the agency has been the atmosphere and the water it contains.

Prior to the systematic study of geology, it was a continuing source of wonderment and superstition that creatures once living should now be found as parts of solid rock, and layers of sea shells found as ridges of high mountains. Today we take these evidences as proof of the dynamic movements of the earth's surface, of erosion and sedimentation, of the evolution of living forms. We piece together the information until we have a pattern that fits all the data. We can trace gradual change in the fossil remains of plants and animals as we study older and older beds, and notice how species converge to simpler common ancestors, until finally we get back to a time when there is no sign of life. We study the sequences of bedded rocks, giving them names and cataloguing them according to the extinct animals and plants that were living at the time of their deposition. When these sedimentary beds become compressed into rocks and are associated with a mountain-building event, we place this event, or *orogeny,* in its correct place in our story.

### Relative Time Scale of the Planet's History

It was a major turning point in earth science when, in 1799, "Strata" Smith in England announced he had discovered that sedimentary strata of the same age consistently showed the same types of fossils. He demonstrated a complete sequence with proof that the succession of fossils was always in the same order. By 1808, French geologists had studied the fossils in the sedimentary beds of the Paris basin sufficiently to be able to correlate them with the strata worked out in England. The worldwide study of historical geology had begun.

These studies led eventually to the theory of evolution and great changes in man's approach to the historical aspects of the universe about him.

No longer was dogmatic creed to be superposed, by force if necessary, on facts denied by much material evidence. Centuries of superstitions fell before the challenging and rapidly advancing hypotheses about all manner of natural phenomena related to the planet earth and its relation to the universe.

In addition, these studies led to extensive interpretations of past advances and recessions of the seas on continental areas. This is natural, since the fossils were largely of creatures that lived in water and had durable parts, such as shells, and that were best preserved in sedimentary materials like muds and silts deposited in shallow seas. Since a considerable proportion of the continental areas is covered with sedimentary rocks, these studies have been of practical importance—particularly since accumulations of petroleum occur in such sedimentary cover rocks. However, had a means of correlation become available, it is probable that volcanic and mountain-building events would have been the more prominent features of the geologic time scale.

A fairly typical cross section of continental rocks is shown in Fig. 11, see also Plate I. This illustrates a gorge of the Colorado River. The crystalline rocks of the so-called *basement complex* represent the roots of an old mountain belt, which was eroded flat before the sea transgressed the land and deposited a layer of sediments. This eroded surface is known as an *unconformity*. The layer of sediments on it was then uplifted and beveled off by erosion, making a second unconformity before

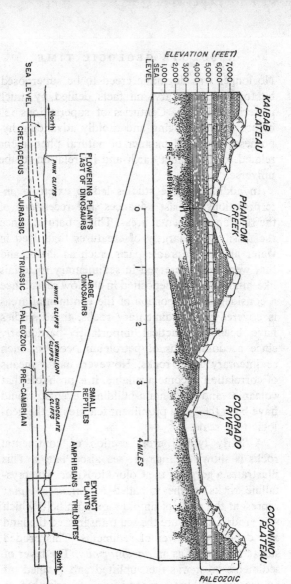

Fig. 11

the sea came in again and deposited the upper beds. Now crystalline rocks, the roots of the old mountain belt may originally have been formed of sedimentary materials, folded and crumpled by compression and buckling of the earth's crust. From the complexity of the structure it can be seen how difficult it is to select which transgression of the sea, or which geologic event, is the important one deserving a place on the area's geologic map. Actually, the material exposed at the surface is mapped, because information on sub-surface material is generally lacking. It is possible that the mountain-building event characterized by the crystalline basement rocks is the most important happening in this area, as it represents a time at which this part of the continent probably first emerged from the ocean depths.

Except in regions where there is no sedimentary cover, a geolgic map rarely shows any but the more recent events. The complexities make it too difficult to do otherwise.

In summary, then, we see earth history expressed at present largely in terms of transgressions of the sea and in the resulting thicknesses of sediments related to the fossil remains of living forms

*Fig. 11. (opposite) The Grand Canyon is so deep because some great force from below lifted the entire region from sea level to more than a mile in altitude. Before this uplift the original land surface, shown as the Pre-Cambrian, was under water and covered by easily eroded sediments. The great depth permits the geologist to study strata and fossils that were marine in origin.*

which were evolving through the geologically more recent past. The greater events in which mountain belts rose in globe-circling chains are left largely as gaps in the sedimentary record, because the method for dating and correlating them was not developed. With the advent of rapid and precise means of age measurement by radioactive methods, this is no longer so; the history of the earth in the future will more truly cover both the great expanse of time involved and the logically more important happenings.

### Principles of Geologic Time Measurement

We have seen that the earth has had a history of large surface events such as the development of systems of mountains and areas of continents. These occurred at different times in different regions. Now only the roots of the mountains are left, owing to erosion in the ages that have elapsed. These mountain systems appear to have added continental areas to the earth's surface at the expanse of the ocean floor, forming stable blocks that do not sink, once they have come into existence.

The measurement of geologic time therefore includes the measurement of the age of the old mountain systems that make up the different parts of the continental crust. So, what we really are measuring is the time that has elapsed since the minerals in granites and other rocks crystallized. Our measurements will reflect one of two events: the time when molten materials came up into the crust, or the time when crustal materials were so deformed and heated that they recrystallized into their existing forms. Also we are able to determine the age of the earth itself (within fairly narrow limits) and the age of the elements that make up our part of the universe, by methods that differ from those used on crustal materials, as we shall see later.

In the later stages of earth history geologists are concerned with measuring the distinct periods in the evolution of living forms and the times when sedimentary sequences were deposited. This is more difficult, and in general we must seek the answers by bracketing the events with other measurements dependent upon the crystallization of some mineral containing a radioactive element.

In a crystalline rock such as a granite (see Plate II) the individual mineral grains are formed at a high temperature. Each crystal is not a pure chemical compound, although the components do come together in approximately the standard proportions for the particular mineral. All mineral crystals contain varying amounts of impurities: atoms of other elements that do not properly belong in the particular crystal structure. Because small amounts of almost all elements occur in all kinds of minerals, it is possible to measure the radioactivity of almost all mineral samples, whether the essential component of the material is a radioactive element or not.

For example, the ordinary mineral grains in a body of granite contain measurable amounts of uranium and thorium. These elements enter the crystal structures during the latter's growth by occasional substitutions for the correct atoms, or by simple inclusion in imperfections of the growing crystal structure.

When a crystal is formed, it is a tight system that sometimes lasts for billions of years without disturbance. On other occasions the system is not tight and certain elements may diffuse through it or exchange with other elements in the environment. Thus, we must be careful in choosing mineral types with which to measure geologic age.

The methods of measuring geologic time by radioactivity require that a parent radioactive element be present in one of these tight crystal structures, and the crystal must have remained a closed system throughout the length of time to be measured. As the radioactive parent element undergoes spontaneous disintegration, it builds up a daughter product which provides a measure of the time since the crystal was mineralized. For example, if a pure crystal of uraninite, $UO_2$, contained no lead at the time of its formation, the breakdown of uranium into lead would start building up lead from zero at a rate depending upon the decay constant of the uranium. If this is known, and the content of uranium in the mineral sample, as well as the content of the lead daughter product, is measured carefully, the time required to build up the measured amount of lead can be calculated. If, on the other hand, the crystal structure contained some lead when the crystal was formed, the increasing amount of lead did not start from zero, and it is not possible to calculate the age unless some other measurements are made. Fortunately, there is a way of doing this. I shall describe it later on.

Since it is not necessary that the mineral chosen for age measurement be one in which the parent element is a major component, it is possible to make age measurements on a mineral such as mica, which has a small content of rubidium as an impurity in the structure. One of the isotopes of rubidium is radioactive $Rb^{87}$, which breaks down to form an isotope of strontium $Sr^{87}$. Again, if there is any common strontium present in the crystal at the time it was formed, which generally is so, a special method must be used to determine how much was there.

## Parent and Daughter Isotopes

The measurement of geologic time, therefore, depends upon finding a crystal structure of a mineral so resistant to change that it will remain essentially intact during its entire lifetime, and it must have a measurable quantity of some radioactive parent element, a measurable quantity of a daughter product forming from the disintegration of this parent, and in some way we must be able to determine the amount of daughter product present originally in the crystal structure. Furthermore, there must be no loss or gain of either the parent or daughter element throughout the life of the crystal, by any process such as diffusion or recrystallization or contamination from the surrounding region. It is truly amazing to think that in a rock such as a granite there are little grains of minerals throughout the entire rock, each of which is a minute clock capable of telling the exact time since the formation of the rock.

Looking into the question in a little more detail, we see that we are dealing with radioactive isotopes rather than elements, and that the daughter products are also isotopes of some other element. For instance, $U^{238}$ breaks down to form an isotope of lead $Pb^{206}$; $U^{235}$ breaks down to form $Pb^{207}$; and $Th^{232}$ forms $Pb^{208}$. Among the other important pairs are $Rb^{87}$ breaking down to form $Sr^{87}$, and $K^{40}$ (potassium) forming $A^{40}$ (argon). Although normally the relative abundances of isotopes are constant, we find that among radiogenic isotopes (those isotopes that are formed from the breakdown of some radioactive parent) there is a considerable change in relative abundance.

For example, the isotopic abundance of common lead is approximately as follows:

| | |
|---|---|
| $Pb^{204}$ | 1.4% |
| $Pb^{206}$ | 26 % |
| $Pb^{207}$ | 21 % |
| $Pb^{208}$ | 52 % |

If, however, we measure the isotopic abundance of lead in an old crystal of uraninite, we find a much greater amount of $Pb^{206}$ present; in fact, if there were no lead at all in the crystal structure to start with, all the Pb would be the 206 and 207 isotopes derived from the uranium isotopes present. Therefore, the relative abundance of radio-

Plate 1. A river can cut into rock only when there is a gravitational gradient to drive the water; thus, the steeper the terrain, the greater the power of the water to erode the rock. The Grand Canyon resulted from the uplift of a plateau originally at sea level to a great elevation in a relatively short time. The time since the uplift has been insufficient to develop the rounded forms of most topography. (Photograph by Paul Caponigro)

Plate 2. A slice of granite no thicker than paper between two pieces of Polaroid looks like this under the microscope. The rock is made of crystals of various minerals.

Plate 3. The geochronology laboratory at the Massachusetts Institute of Technology. (Photograph by R. Paul Larkin)

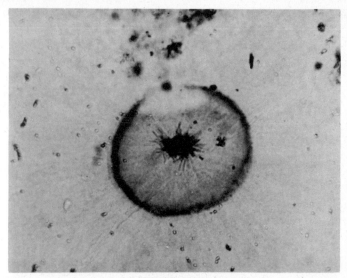

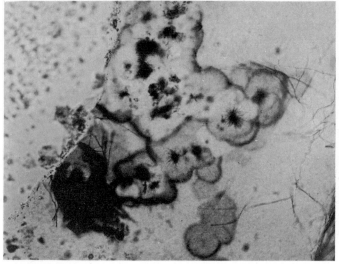

Plate 4. These are microscope photographs of the oldest known remains of living forms. They resemble some species of modern Algae. They have been studied by E. S. Barghoorn, of Harvard University, and S. A. Tyler, of the University of Wisconsin, and dated as 1600 million years old by the author. (Photograph courtesy of Professor Barghoorn)

Plate 5. The story of evolution has grown from the study of the fossilized remains of living organisms preserved in rocks. This picture shows *Paradoxides bohemicus*, a trilobite. (Photograph by R. Paul Larkin)

Plate 6. The moon is believed to have formed at the same time as the earth as a separate condensation nucleus. Composed of the same material, it would have heated by radioactivity enough to cause melting in the interior, and therefore probably had volcanic eruptions. The eruptions may have caused many of its surface features. (Photograph by Harold M. Lambert)

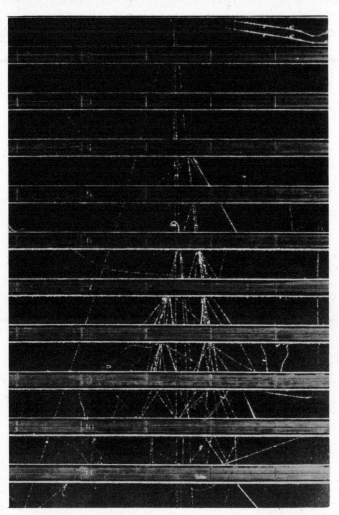

Plate 7. We cannot escape the radiations caused by the high velocity particles that rain on our atmosphere and are known as cosmic rays. As the tracks in the picture show, they can penetrate a considerable thickness of brass. Cosmic ray damage represents about a quarter of the total radiation damage we suffer continuously from all natural sources, exclusive of man-made sources. (Courtesy of Bruno Rossi and R. B. Brode of M. I. T.)

Plate 8. Half Dome is an example of giant processes at work. The battle between uplift and erosion stages a mighty scene at Yosemite. (Photograph by Paul Caponigro)

genic isotopes can vary considerably, depending whether or not these isotopes have been in close association with the parent element over a long period of time.

Thus it can be seen that there should be a normal slow change of the abundance of radiogenic isotopes with time in the earth's crust corresponding to the normal average slow breakdown of the total amount of parent in the crust. For example, surface rocks are estimated to contain about 0.035 per cent of rubidium and about 0.022 per cent strontium. The isotopic abundance in each element may be shown in the following way (Fig. 12). The radioactive rubidium 87 isotope breaks down at a rate such that half of it would have gone in 50 billion years, or about 5 per cent in the lifetime of the earth. Its breakdown product is strontium 87. Because rubidium 87 appears to be six times more abundant than strontium 87 in the earth's crust, the strontium 87, by the addition of the radiogenic material, should have increased by 30 per cent during the lifetime of the earth.

Actually, measurement of very old strontium minerals which do not contain much rubidium has shown this not to be so. The change is much smaller. The explanation is this: Rubidium is greatly enriched in the crustal rocks, and the actual amount in the upper mantle is very much less than the amount of strontium. This is interesting, because it provides us with another way of studying the relative enrichment of elements in the upper mantle and crust.

In a similar manner, we gain information about

the origin of the atmosphere by considering the enrichment of radiogenic argon 40 in it. The argon 40 in the present atmosphere is equal to about 25 per cent of the estimated total argon 40 produced by the breakdown of K⁴⁰ in the earth since its beginning. The gases released during the development of the continents, which cover only about 20 per cent of the earth's surface, plus oceanic volcanic islands, would completely account for this argon

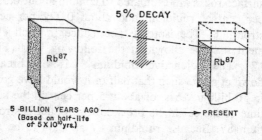

**5% DECAY**

Rb⁸⁷        Rb⁸⁷

**5 BILLION YEARS AGO** ──────────▶ PRESENT
(Based on half-life
of 5 X 10¹⁰ yrs.)

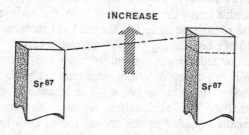

**INCREASE**

Sr⁸⁷        Sr⁸⁷

*Fig. 12. Radioactive nuclides decay at known rates. Knowing how much of a given nuclide—rubidium 87, for instance—a mineral has now and its decay rate, we can calculate how much there was at any time in the past. Similarly, because rubidium 87 breaks down to form strontium 87, its change also can be calculated.*

40. However, this rules out the possibility of much escape of gases from the mantle 4.5 billion years ago, or at the time the core and mantle separated.

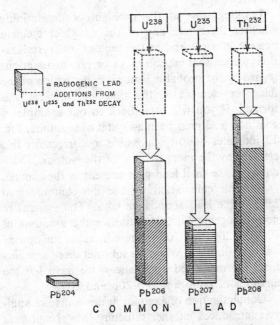

*Fig. 13. The standard relative abundances of the isotopes of common lead are shown above. As lead 204 is not changing with time, the standard ratios of common lead isotopes 206, 207, and 208 relative to lead 204 can be subtracted from any measured isotope distribution found in a uranium or thorium-rich mineral to give the true amount of radiogenic lead 206, 207, or 208 formed from the breakdown of the uranium or thorium. This will correct for common lead present in the mineral at the start.*

83

This is evidence in support of the hypothesis that the oceans and atmosphere increased slowly, keeping pace with the growth of the continental crustal material.

Looking now at the problem of determining whether there was any of the daughter element present in a mineral at the time of its crystallization, we see that we can make certain assumptions regarding the probable isotopic composition of the daughter element at the time the mineral was formed. If you will refer back to our example of common lead present in a crystal of uraninite, Fig. 13 indicates how it is possible to correct for this element by the measurement of the isotopic abundance of the final lead now present in the sample.

For this calculation we use the common lead ratio of the lead abundance table. The amount of non-radiogenic Pb 204 indicates the amounts of Pb-206, 207, and 208 in the mineral at the time of its crystallization. When we subtract these amounts from the measured abundances, we have left the amounts of radiogenic Pb-206 and 207.

Contamination of samples in the course of analysis introduces another difficulty into the process of determining geologic age through radioactivity measurements, but we use refined methods of correcting for the contamination. A little further on I shall go into this in some detail, and with some simple mathematics, because a concrete example will illuminate the technique of age measurement and at the same time show the relationship between certain forces involved in these complex experiments and certain forces studied in simple experiments performed in every high school physics laboratory.

The accuracy attained in these contamination corrections is truly amazing. Let us suppose, for example, that we have a young mineral containing argon 40 developed in the breakdown of potassium 40. Although each gram of the mineral sample contains only $10^{-5}$ cubic centimeters of argon 40 under standard conditions of temperature and atmospheric pressure, it is possible to separate the argon from the sample, to purify it and measure it. But the amount is so small that any air contamination remaining in the sample or any air leak in the course of analysis would destroy the accuracy because all samples of air contain a small amount of argon 40. A correction is necessary. Fortunately, the air also contains the argon isotope $A^{36}$ in such proportion that the ratio of $A^{36}$ to $A^{40}$ equals .00337. So, if we measure the amount of $A^{36}$, we can calculate the amount of $A^{40}$ that is air contamination. When we subtract that amount from the total argon 40 in the mineral sample, we can determine the true amount of radiogenic $A^{40}$ in the sample, despite the slight $A^{40}$ contamination from the air. In very young materials it is almost impossible to eliminate all contamination; to obtain accurate age measurements, the correction is essential.

## History of Age Measurements

But before we go into the techniques in detail, let us review briefly the history of this fascinating branch of science. It is important because the development of accurate geologic time measurement coincides with the growth of our knowledge of radioactivity.

It all started, quite by chance, when the French physicist Henri Becquerel (1852–1908) left a photographic plate and a piece of a uranium salt together in a drawer. The next time he looked, something had darkened the plate; in other words, it had been exposed to some unknown radiation. This fortuitous incident turned Becquerel and his contemporary experimenters onto a new track that led ultimately to grand new achievements in physical science: our knowledge of the atomic nucleus, radioactive disintegration, nuclear fission, and fusion.

Soon after Becquerel's discovery, Pierre and Marie Curie began their long and now famous search for sources of radioactivity and eventually succeeded in isolating radium and several other radioactive elements. The Radium Institute in Vienna and other organizations joined in the search for radioactivity in other elements, and it has continued to this day.

It now is known that all atoms with Z-number above 82—that is, whose nuclei contain 82 or more protons—are radioactive, and many naturally occurring lighter ones are also. The heavy elements were found to be related to three natural radioactive series. These are the uranium series, the thorium series, and the actinium series. The uranium series is shown in Fig. 9. The other two series are of a similar nature. In all, there are about sixty known naturally occurring radioactive nuclides. Most are short-lived and exist only because the breakdown of some radioactive parent element or some other nuclear reaction forms them continuously.

The application of research in radioactivity to understanding of the earth and universe had its beginning shortly after the turn of the century. Much had been learned about the decay scheme of uranium, the constancy of the ratio of uranium to radium, and it had been demonstrated that lead was the stable end product of the disintegration series. It soon became apparent that the radioactive process involved the generation of heat and thus was of great significance in the study of earth history.

In his book *Radioactivity in Geology* (1909), John Joly opened the way to new understanding of earth heat and the effect of radioactivity on the time required for the earth to cool from its formation temperature. The nineteenth-century authority on thermodynamics, Lord Kelvin of Glasgow University (1824–1907), had studied the cooling question and come to the conclusion that the earth was only 24 million years old. But Joly demonstrated that radioactivity in the rocks changed the picture altogether. (He also was the first to recognize the possibility that radiogenic heat—heat generated in radioactive disintegrations —supplied the energy for the great mountain-building processes.)

Another discovery was to be of much interest in the question of earth age. This was that helium, first found on the sun and then on earth, was a product of the breakdown of uranium and thorium. It was another of the tools that, early in this century, enabled scientists to make their initial attempts to measure earth age through the process of radioactivity. The great Lord Rutherford was the first to study the relation of helium to uranium in minerals, but the helium loss in the highly radioactive minerals he measured was so large his results indicated very low ages.

After Lord Rutherford's work with helium came systematic investigation of lead-uranium ratios, and this research quickly created a completely new dimension in geology—the approximate measurement of geologic time in absolute terms. The period of intensive study of lead-uranium and lead-thorium ratios in radioactive minerals of various ages had reached a climax in 1931 with the publication of *The Age of the Earth,* by Knopf, Schuchert, Kovarik, Holmes, and Brown, in the National Research Council's "Physics of the Earth" series.

In brief, this line of investigation made two significant findings: that a fairly orderly progression of ages back to about 2 billion years in various parts of the earth could be demonstrated, and that this progression was in agreement with the geological interpretation of the succession of events. The radioactivity studies corroborated the older interpretations which showed that certain mountain belts were older than other belts cutting through them. But the data were sketchy, yielding only a few points on which to hang all geological history, and there were many *discordances,* or failures of independent lines of investigation to produce results in agreement with each other. What caused most of the trouble was the presence of common lead that had existed in the minerals at the time they were formed. A new approach was needed.

More powerful new methods were found, in the period 1935 to 1940, in the study of isotopes and development of mass spectometry, which is a technique for determining the masses of positively charged particles and for finding the relative amounts of the isotopes in an element. In classic papers published from 1938 to 1941, A. O. Nier, then at Harvard University, opened up the possibility of making precise age determinations by separate isotope measurements and by making corrections for the presence of daughter elements at the time of crystallization of a mineral. Gains or losses of the elements of interest could be detected by means of *concordancy* of different isotopic ratios in the same sample.

This period also saw the first publications on artificially produced radioactivity, and the uses of this tool in geological science advanced rapidly. In 1935, Nier demonstrated the existence of the radioactive isotope $K^{40}$. About the same time, in Berlin, a team of four scientists—Lise Meitner, Otto Hahn, Fritz Strassman and Otto Frisch—was carrying out a series of perplexing experiments that led eventually to the discovery of nuclear fission. The age of experimental nuclear physics had begun.

It was in the same period that Charles S. Piggott and William D. Urry were studying the deposition of ionium, a daughter product in the breakdown of $U^{238}$, in ocean-floor sediments. It was found that sea water deposited ionium in excess of its equilibrium amount of parent element uranium; as subsequent deposition of sediments buried the ionium, it decayed with an 84,000-year half-life. Measurements taken in the topmost few centimeters of ocean-floor sediment showed that the radioactivity gradually diminished with depth. Under ideal conditions, this could be interpreted to give the rate of deposition of sediments, a measurement of considerable value to the understanding of the geology of these vast areas, and to the investigation of the abundance of elements in the earth's crust.

After studying under Nier, at the University of Minnesota, L. T. Aldrich, of the Carnegie Institution, became interested in geochronology, and, together with G. W. Whetherill, at the Department of Terrestrial Magnetism of the Carnegie Institution, led the important investigations of the ratios of argon 40/potassium 40 and strontium 87/rubidium 87 in minerals. Since Nier's original work great strides have been made in the study of radiogenic isotopes of lead and their variations, both in measuring the age of the earth and meteorites and in seeking better understanding of the relationship between the continental crustal rocks and the mantle below them. And of major importance in archaeology and the geology of "recent" events— a few tens of thousands of years ago at the most—

has been the development of the carbon 14 method of analysis by W. F. Libby and his associates at the University of Chicago.

Many other avenues of study in the fields of age measurement and radioactivity have been opened up, but it is beyond the scope of this book to consider them all in detail. In this brief survey of the progress of the investigations we have tried to present a broad picture of the problems involved and to convey a general idea of what kinds of events are being measured. Now we are ready to get down to specifics.

## Techniques and Instruments

First, we must examine the physical principles underlying the measurements needed in age determination. As we have seen, it is necessary to measure the absolute amount of a single isotope in a sample of mineral or rock. Suppose, for example, that we are measuring the $A^{40}/K^{40}$ ratio in a sample of black mica (biotite) from a granitic region in the Northern Territories of Australia.

The mica minerals are particularly good for this method of age determination because they contain much potassium and are easily concentrated from the rock. Geologists collect a sample of the granite in the field and take it to the laboratory, where it is crushed and ground and the mica separated from it. A sample of the mica is then analyzed chemically (or by flame photometer or other standard procedure) for potassium. For refined work this measurement is made by mass spectrometer. Analysis gives the total amount of potassium in the sample, and since the isotope $K^{40}$ bears a constant relationship with the other isotopes of potassium, the amount of $K^{40}$ can be calculated. Actually, the proportion of $K^{40}$ in total K is .000119 per cent.

Another sample of the mica is then weighed and placed in a furnace which is part of a vacuum-tight system such as shown in Plate III. Almost all traces of air contamination are removed from the system and from the sample by warming the sample and pumping out all of the gases for a period of time. The furnace is then heated to such a temperature that the sample of mica melts and the argon which has formed from the decay of $K^{40}$ is released into the system. If no other argon from any air contamination were present, this argon would all be radiogenic $A^{40}$. But if there are traces of air contamination still remaining either in the sample or in the system, there will be a small amount of $A^{36}$ which always accompanies air argon in a ratio of about one atom in 300.

In order to measure the absolute amount of $A^{40}$ released from the sample, it is necessary to add a known amount of another isotope of argon against which the abundance of $A^{40}$ can be compared. Generally $A^{38}$ is added. (An isotope used for this purpose is known as a "spike.") A known number of atoms of $A^{38}$ is metered into the apparatus through a system in which the volume can be measured very accurately. Most of the gases from the sample are then removed from the system with a material such as hot titanium metal, which absorbs most gases readily. Hydrogen is removed by oxidizing it in the presence of heated copper oxide and then condensing the water that is formed at liquid nitrogen temperature, along with any other gases that may condense at the same temperature. Almost pure argon is left in the system, and this is now composed of three different quantities. First there is the radiogenic $A^{40}$; next there possibly is a small amount of argon from air contamination ($A^{40}$, $A^{38}$ and $A^{36}$); and thirdly there is the known amount of $A^{38}$ spike. This mixture of argon isotopes is now ready for measurement in a mass spectrometer, which determines the ratio of the isotopes 40, 38, and 36.

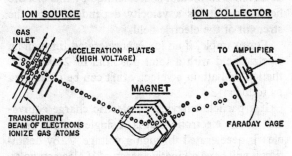

*Fig. 14. An ion of mass $m_1$ and electric charge e will be separated from another ion of mass $m_2$ and the same charge e as it passes through a magnetic field, if traveling the same initial path. This is the principle of the mass spectrometer.*

A diagram of a mass spectrometer is shown in Fig. 14. The argon is admitted to the instrument through a small opening; commonly only about $10^{13}$ atoms pass through this opening per second

by diffusion into the evacuated part of the spectrometer. As the individual atoms of argon enter
into the mass spectrometer, they meet a transcur-
rent beam of electrons generated by a hot filament
and pulled across their path by an electric field.
Most of the atoms of argon are ionized by the
electrons; this means that when an electron strikes
a neutral atom of argon it removes one of the
atom's outer electrons, leaving it with a single net
positive charge. The argon atom, then known as a
positive ion, can be moved in space by placing it
in an electric field. This electric field is present in
the volume where the gas is being ionized; as soon
as an ion is formed, it is accelerated by this electric
field and obtains a velocity dependent upon the
strength of the electric field.

For example, if an ion is accelerated through an
electric field with a total potential of 2000 volts,
the ion will attain a velocity that can be calculated
as follows:

If a particle carrying an electric charge equivalent to one electron (that is, a single ionized particle) is accelerated through a voltage V, by defini-
tion it will have a kinetic energy of V electron volts.
The electron volt has been determined through experiment to be equivalent to $1.60 \times 10^{-12}$ erg. In
this example, therefore, the kinetic energy imparted
to the argon 40 ion is $2,000 \times 1.60 \times 10^{-12}$ erg.

Since we know that kinetic energy equals one
half the mass ($m$) times the square of the velocity
($v$), or

$$K.E. = \frac{1}{2}mv^2$$

we can proceed to solve the equation for $v$ after we have calculated the mass of the argon 40 ion. To do this, we must go back to one of the milestones in the progress of physical science, to an hypothesis advanced by Amadeo Avogadro (1776–1856). One of Avogadro's conclusions was that there are $6.02 \times 10^{23}$ atoms in an amount of any element equal to its atomic weight in grams. The atomic weight of $A^{40}$ is 40; so 40 grams of $A^{40}$ will contain $6.02 \times 10^{23}$ atoms, or one atom of $A^{40}$ will weigh $\dfrac{40}{6.02 \times 10^{23}}$ grams. (The constant $6.02 \times 10^{23}$ is known as Avogadro's Number and applies also to the number of molecules in a weight of a substance equal to its molecular weight.)

So, substituting in our K.E. equation, we have

$$2000 \times 1.60 \times 10^{-12} = \frac{1}{2} \times \frac{40}{6.02} \times 10^{-23} \times v^2$$

or

$$v = 0.98 \times 10^7 \text{ cm/sec}$$

or

$$= 216,000 \text{ miles per hour}$$

The velocity of an argon 38 ion, being of lower mass, is greater. Because

$$v^2 = \frac{2(\text{K.E.})}{m}$$

The ratio $v_1^2 / v_2^2$ for two different ions is inversely proportioned to their masses

or

$$v_1^2 / v_2^2 = m_2 / m_1$$

95

so that

$$v_1/v_2 = \sqrt{m_2/m_1} \text{ or } v_2 = v_1\sqrt{m_1/m_2}$$

Therefore the argon 38 ion would be traveling

$$0.98 \times 10^7 \times \sqrt{40/38} \text{ cm/sec}$$

or

$$1.00 \times 10^7 \text{ cm/sec}$$

A particle of mass $m$, with an electric charge $e$, moving at right angles to a magnetic field of strength $H$, will have a force acting on it equal to $Hev$, and it will cause the particle to travel in a circular path. Now we know from the mechanics of a weight on a string that for a circular motion in a path of radius $r$, there must be a force equal to $mv^2/r$ acting on the mass toward the center of the circle.

Thus in the magnetic field the particle will take a circular path of such a radius that the magnetic force $Hev$ equals $mv^2/r$, or the radius $r = \dfrac{mv}{He}$.

Again referring to our two ions, argon 40 and argon 38, which we will designate by subscripts 1 and 2, we see that

$$r_1/r_2 = m_1v_1/m_2v_2$$

But

$$v_1/v_2 = \sqrt{m_2/m_1}$$

so that

$$r_1/r_2 = \sqrt{m_1/m_2}$$

In other words, if the two ions are accelerated through the same voltage and pass through the same magnetic field, they will be deflected at different angles, because they will travel a short distance along circular paths of different radii as shown in Fig. 14.

Large numbers of these ions traveling along these paths are called ion currents, or ion beams. It can be seen that if the ion beam is measured when it enters a fixed slit beyond the magnetic field, a variation of either the accelerating voltage or the magnetic field will cause one or the other of the types of argon ion to be measured. This is actually done in practice, and the ion beams for argon 38 and argon 40 are alternately measured as they are moved back and forth across the pick-up slit by changing the magnetic field slowly. Thus the ratio of argon 38 atoms to argon 40 atoms entering the mass spectrometer can be determined accurately.

An example of a record showing the amplified current from ion beams of $A^{38}$ and $A^{40}$ taken during an age measurement is given in Fig. 15. The peaks get progressively smaller because the sample is being used up. In a normal analysis $A^{36}$ is also measured to indicate possible air leaks or contamination.

Returning now to our example of an age measurement of mica from Australia, we see how it is possible to determine its age using the isotope dilution technique. Using an actual measurement as an example, the gas from a 1.81-gram sample was "spiked" with $10.95 \times 10^{15}$ atoms of argon 38 introduced into the furnace from a measured volume.

The total gas was purified and the ratio of argon isotopes measured on a mass spectrometer. The measured ratio of argon 36 to argon 40 was found to be negligibly small so that it could be safely assumed that there had been no air contamination.

The measured ratio of argon 38 to radiogenic argon 40 was found to be 0.446, so that the actual

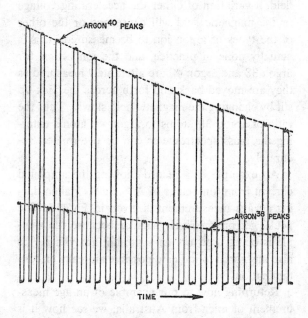

*Fig. 15. The relative amounts of argon 40 and argon 38 are shown in this mass spectrometer record. Argon 40 is indicated by the taller peaks and argon 38 by the lower peaks. Both amounts decrease in time as the original sample is used up, but the proportional heights remain the same.*

number of argon 40 atoms in each gram of sample
was given by

$$\text{Argon 40 atoms/gm} = \frac{10.95 \times 10^{15}}{.446 \times 1.81} = 13.58 \times 10^{15}$$

The total potassium was found to be 4.21 per cent
by separate analysis. But of this total potassium
only .0119 per cent of the atoms are radioactive
potassium 40, or in one gram of sample there were

$$.0421 \times .000119 \times \frac{\text{Avogadro's Number}}{\text{atomic weight}}$$

or

$$77.1 \times 10^{15} \text{ atoms of potassium 40}$$

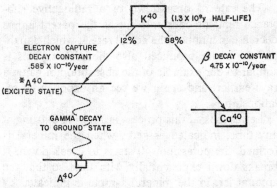

*Fig. 16. Potassium 40 breaks down to calcium 40 by
beta decay in 88 per cent of the cases, and to argon
40 by electron capture in 12 per cent of the cases,
with the disintegration constants shown. This ratio
is a statistical constant and does not change with time.
The diagram also shows that argon 40 forms first in
an excited state ($*A^{40}$), which then transforms to argon 40 in the ground state by emission of a gamma ray.*

Therefore, $13.58 \times 10^{15}$ atoms of argon 40 had developed from the radioactive breakdown of some number of potassium 40 atoms, of which there are now $77.1 \times 10^{15}$ atoms left.

Potassium 40 breaks down to calcium 40 also (see Fig. 16) in a ratio of about 8.47 calcium atoms to one argon atom. This means that at the time of crystallization of the mica, each gram of it contained potassium 40 atoms equal to

$$(77.1 \times 10^{15}) + (13.58 \times 10^{15}) + (8.47 \times 13.58 \times 10^{15})$$

or

$$205.6 \times 10^{15} \text{ atoms.}$$

The rate of breakdown of a radioactive substance remains constant, but as the parent atoms decrease in number, the actual rate at which daughter atoms are produced also diminishes. This is similar to our example of the absorption of gamma rays earlier, and again we see an exponential relationship.

Let us look at this process in Fig. 17. Starting at any time, indicated as zero, when the mineral is formed, the potassium 40 starts to break down. It breaks down exponentially with time so that the amount left in the mineral sample is indicated by the curve marked $K^{40}$. In order to test this we can see from the curve that one half of the $K^{40}$ is left after $1.31 \times 10^9$ years, which is the half-life of $K^{40}$.

In the same sample, however, the products of the breakdown, $Ca^{40}$ and $A^{40}$, start building up from zero, and the total number of atoms of both combined equal the number of atoms of $K^{40}$ that

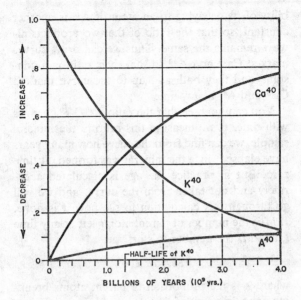

*Fig. 17. A unit amount of potassium 40 breaks down exponentially with time with a half-life of $1.31 \times 10^9$ years, as seen on the curve marked $K^{40}$. If this occurs in a closed system, the decay products $Ca^{40}$ and $A^{40}$ will start to build up in a ratio shown by the two other curves. The sum of these two nuclides at any time equals the amount of $K^{40}$ that has disintegrated.*

have gone. Therefore, the sum of the curves marked $K^{40}$ and $Ca^{40} + A^{40}$ always equals the original number of atoms of $K^{40}$ existing at time zero.

Now we know that $A^{40}$ and $Ca^{40}$ always form in the same ratio. In other words, the probability of a $K^{40}$ atom's decaying to an atom of $Ca^{40}$ in a unit of time is a constant, and similarly the probability of its decaying to an atom of $A^{40}$ is another constant, so that the ratio of the two products always remains the same. Thus we can draw curves marked $Ca^{40}$ and $A^{40}$ so that their ratio is a constant, and they both add up to the curve marked $Ca^{40}$ and $A^{40}$ at any time.

You can now see that the ratio $A^{40}/K^{40}$ changes with time. If we measure this ratio in the mineral sample, we can find from the curve how many years have elapsed since the mineral was formed, at time zero. But in practice the age is calculated accurately and not taken from the curve, and we will go through this calculation to see how it is done. If $M$ is the number of parent atoms left at any time $t$, and $M_0$ the number to start with

$$M = M_0 e^{-\lambda t}$$

where $\lambda$ is the proportion of parent atoms breaking down in a unit of time.

For potassium 40, the decay constant $\lambda$ is $5.28 \times 10^{-10}$ per year. In our example $M_0 = 205.6 \times 10^{15}$ atoms and $M = 77.1 \times 10^{15}$ atoms, so that we can solve for $t$, the time required for this much decay of the potassium. This can be done by taking the common logarithm of the equation

$$\log M = \log M_0 - \lambda t \log e$$

or

$$16 \log 7.71 = 17 \log 2.05 - 5.28 \times 10^{-10} t \times \log e$$

As $\log e = .434$ the solution gives

$$t = 1860 \times 10^6 \text{ years, or the mica is 1860}$$
$$\text{million years old.}$$

In the measurement of $Sr^{87}$ and $Rb^{87}$ there is no satisfactory way to have these isotopes in the form of a gas, like argon. Samples of compounds of strontium and rubidium must be heated on a filament in the mass spectrometer. Ions of strontium or rubidium are emitted by thermal ionization, rather than created by electron bombardment. These ions are accelerated in the way we have described, and the measurement is otherwise the same. This type of instrument is known as a solid-source mass spectrometer. There are always small corrections to be made in these kinds of measurements; for example, the argon isotopes, because of their different masses, do not flow at the same rate through the small opening, and the strontium atoms come off the filament at a different rate according to their different masses. Generally we can calculate, or correct for, these "fractionation" effects.

### Present-Day Methods

Different investigators have their own favorite methods of measuring geologic age, and all the methods have a useful place. At the present time it appears that two methods are emerging as most generally useful for most of the earth's history. These are the $Sr^{87}/Rb^{87}$ and $A^{40}/K^{40}$ ratios. In general these methods can be used on potassium-rich minerals, which are very common. Although the Sr/Rb method does not depend upon potassium, rubidium commonly is in minerals that are rich in potassium, and it is possible to compare results on the same mineral by the two different methods. Many common rock minerals, particularly the group of micas such as muscovite, biotite, and lepidolite are used. The potassium feldspar, orthoclase, may be used in certain circumstances.

Orthoclase does not yield satisfactory ages in the A/K measurement method, the results generally being about 25 per cent low. The cause of this discrepancy is not yet definitely known. The mineral seems to be usable for Sr/Rb ratios, however. This means that most crystalline rocks such as granites, gneisses, schists, and slates can be used in age determination because they contain satisfactory minerals. Of course, if rocks have been recrystallized, the age that is measured is only the age of the recrystallization.

Applying the A/K method and the Sr/Rb method to the same mineral gives a cross-check useful in determining whether the material is suitable for age measurement. If the two ratios agree, it is fairly good evidence that the daughter products have not been lost since the time of the crystallization of the mineral, because it is to be expected that such dissimilar atoms as A and Sr would be lost at different rates, if any leakage occurred.

One of the most difficult problems in the development of these methods has been the accurate determination of the half-life of the parent element. The values for the decay constants are still not settled, although temporary values are being used. However, the agreement in the ages determined for the same minerals by the different methods indicate that the half-life values must be nearly correct.

Still of major usefulness is the measurement of lead-uranium ratios in minerals that have a high content of uranium. Today in this method it is necessary that the three different age ratios $Pb^{206}/U^{238}$, $Pb^{207}/U^{235}$, and $Pb^{207}/Pb^{206}$, give ages that are in agreement before the age measurement as a whole can be considered acceptable. When this is the case, the age is said to be "concordant." Frequently a separate ratio can be obtained for the same mineral if there is a small amount of thorium present: namely, the ratio $Pb^{208}/Th^{232}$. All these separate measurements give information regarding the history of the mineral and its general acceptability. When a mineral has yielded a concordant age, this age is generally considered to be highly foolproof. These concordant ages have therefore been used to a large extent as key age points against which other methods can be checked. The drawback with the method is the lack of general distribution of radioactive minerals and the fact that these radioactive minerals have suffered such a large amount of radiation damage that they frequently show ages that are not concordant.

## The Carbon 14 Method

Carbon 14 emits a beta particle with an energy of .15 Mev and has a half-life of about 5600 years. It can be created in the neutron pile from nitrogen 14 with thermal neutrons. In the atmosphere the neutrons are liberated by cosmic radiation and are most abundant above 30,000 feet. Most of the neutrons liberated by cosmic radiation eventually go into the formation of carbon 14 out of the nitrogen of the atmosphere, so that carbon 14 production, and also the rate of $C^{14}$ decay, is almost equal to the rate of production of neutrons. The carbon 14 rapidly oxidizes to $CO_2$ and enters the carbon cycle in the surface of the earth. It is estimated that the total carbon in the atmosphere, oceans, and in the biological material on the earth's surface amounts to approximately 8.3 grams per square centimeter of surface.

If in the atmosphere 2.4 neutrons on the average are produced per square centimeter of earth's surface per second, the disintegration rate of the total carbon 14 should be $2.4 \times 60/8.3 = 17.2$ disintegrations per gram of carbon per minute. Actual measurements of the radioactivity of ordinary living carbon on the earth's surface today yield a mean value of 16.1 disintegrations per minute, which is in excellent agreement with the theoretical prediction.

When carbon in some compound becomes buried or removed from the living cycle of carbon, so that no further additions of carbon 14 enter into the compound, the carbon 14 that the compound started with begins to decay on a half-life of 5600 years. For example, if a piece of wood in the tomb of one of the Pharaohs of Egypt was obtained from a tree living at the time, and since that time had remained buried below the surface and out of the range of neutrons produced by cosmic rays, it would have measurably less carbon 14 than a piece of wood living today.

As we have said, this method developed by Libby and his co-workers has proven to be highly successful in archaeological research. In order to demonstrate that the method worked, as well as

AGE DETERMINATIONS ON SAMPLES OF KNOWN AGE

| SAMPLE | SPECIFIC ACTIVITY (cpm/g of carbon) | | AGE (years) | |
|---|---|---|---|---|
| | FOUND | EXPECTED | FOUND | EXPECTED |
| TREE RING | 10.99±0.15 | 10.65 | 1100±150 | 1372±50 |
| PTOLEMY | 9.50±0.45 | 9.67 | 2300±450 | 2149±150 |
| TAYINAT | 9.18±0.18 | 9.10 | 2600±150 | 2624±50 |
| REDWOOD | 8.68±0.17 | 8.78 | 3005±165 | 2928±52 |
| SESOSTRIS | 7.97±0.30 | 7.90 | 3700±400 | 3792±50 |
| ZOSER: SNEFERU | | 7.15 | 4750±250 | |
| ZOSER | 7.88±0.74 | | | 4650±75 |
| | 7.36±0.53 | | | |
| SNEFERU | 7.04±0.20 | | | 4600±75 |

*This classic set of measurements of the carbon 14 content of ancient pieces of wood of known age launched an important new method of dating the geologically recent past. Willard F. Libby, of the University of Chicago, developed the method; his associate, James R. Arnold, made these measurements. The objects included a tree trunk from an Arizona excavation, part of a coffin of the Ptolemaic period of Egypt, wood from a floor at Tayinat, Syria, deck board from the funeral boat of the Egyptian Sesostris III, and pieces of wood from the tombs of Sneferu of Meydum and Zoser of Sakkara, Egypt.*

107

to calibrate it, the investigators obtained samples of carbon-bearing material from locations in which the age of burial was known fairly accurately. In this way the method could be tested. The preceding table shows the original set of determinations.

Since this work large numbers of analyses have been made of carbonate-bearing materials of all kinds—including shells from the ocean, carbon in deep ocean waters to determine the rate of circulation, peat in peat bogs, ancient relics of all types, and many things which have both geological and archaeological interest.

The actual measurement of the sample is not too difficult, except for the problem of reducing the background of radioactivity to a low enough level so that the weak beta activity of carbon 14 may be measured accurately. Solid or gas counting may be used; the sample may be a solid coating of carbon black inside a Geiger counter, or the carbon may be in a gas, such as carbon dioxide, when it is introduced into the counter. The counting chamber is surrounded by an iron or mercury shield, to remove the soft component of background gamma radiation, and a ring of anti-coincidence counters which correct for stray cosmic ray activity. These techniques usually reduce the background of the counter to about three counts per minute or so. Without rather special developmental work, it is not generally practicable to measure ages in excess of about 20,000 years, because the radioactivity of the carbon becomes so slight that it is difficult to get an accurate measurement above background radioactivity.

The carbon 14 method does not depend upon the presence of a long-lived radioactive isotope that has remained in existence since the origin of the elements. It depends upon the continuous formation of a short-life radioactive element, by what is believed to be a fairly constant nuclear process. Obviously, the method depends upon constant neutron flux into the upper part of the atmosphere or a fairly predictable quantity of carbon in the reservoir that makes up the cycle. Actually, a correction must be made for the increase in carbon in this carbon cycle during the last fifty years because of the large amounts of coal and oil that have been burned since the Industrial Revolution.

# CHAPTER V

# Memorable Dates in Earth History

Being egocentric creatures, we human beings are understandably inclined to start our examination of the universe by looking at ourselves, at the ground we walk on, at the mountains that tower above us, at the oceans that lap night and day upon our shores and spread their waters farther than the eye can see. Our "universe," therefore, tends to be a local universe, which we soon find, did not start all at the same time. We now must study each part of our environment separately before we can speak of the "age of the universe" with understanding. Later, we get back to the origins of the earth, our galaxy, and the elements of which these are made and, ultimately, to the universe of galaxies.

In this and the final chapter we shall try to set forth the best present answers to these tremendous matters. By stages we shall go from the time the earth's crust became stable—from the first fossil records of the earliest forms of life to the evolution of Homo sapiens—Man. Then, having established a platform and Man to stand on it, we shall let him look upward to ponder the greater problems of genesis. An increasingly well established advance base in this pioneer field of science is our knowledge of the age and origin of the earth.

So far measurement of the oldest rocks of the continental masses over most of the earth's surface has produced an unusually consistent set of numbers. The oldest age measured is approximately 2600 million years, and this value has been found in North and South America, Asia, Africa, and Australia. No older reliable age has been recorded in any other region.

Thus, geologists assume that the continental earth's crust was either too unstable before this time to have been preserved or else had not yet started to segregate from the mantle.

These earlier parts of the continents were probably surrounded by oceans not quite as deep as today's. We believe that these continental areas were not eroded down to a very much lower level than the present sea level, because, on the average, their surfaces are not submerged platforms but are unexpectedly close to the present sea level. One unexplained feature of interest is that most of the largest gold mines in the world occur in these areas.

Following these first mountains of 2400–2800 million years ago came a succession of mountain-building events lasting to the present. The distribution of these is indicated approximately in Fig. 18. The greatest mountain systems of all time have been the youngest, which have swept in great

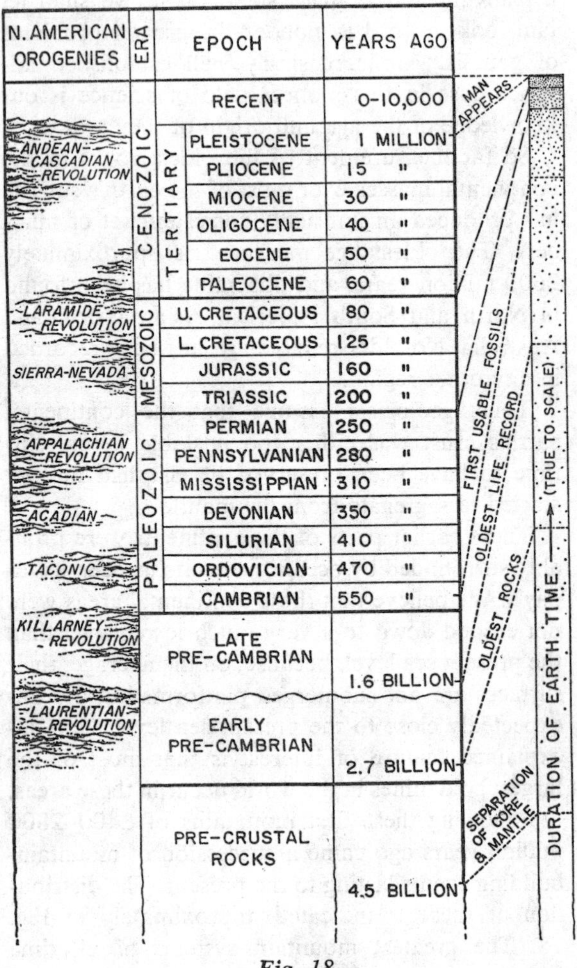

| N. AMERICAN OROGENIES | ERA | EPOCH | | YEARS AGO |
|---|---|---|---|---|
| | | RECENT | | 0-10,000 |
| ANDEAN–CASCADIAN REVOLUTION | CENOZOIC | TERTIARY | PLEISTOCENE | 1 MILLION |
| | | | PLIOCENE | 15 " |
| | | | MIOCENE | 30 " |
| | | | OLIGOCENE | 40 " |
| | | | EOCENE | 50 " |
| | | | PALEOCENE | 60 " |
| LARAMIDE REVOLUTION | MESOZOIC | U. CRETACEOUS | | 80 " |
| | | L. CRETACEOUS | | 125 " |
| SIERRA-NEVADA | | JURASSIC | | 160 " |
| | | TRIASSIC | | 200 " |
| APPALACHIAN REVOLUTION | PALEOZOIC | PERMIAN | | 250 " |
| | | PENNSYLVANIAN | | 280 " |
| | | MISSISSIPPIAN | | 310 " |
| ACADIAN | | DEVONIAN | | 350 " |
| | | SILURIAN | | 410 " |
| TACONIC | | ORDOVICIAN | | 470 " |
| | | CAMBRIAN | | 550 " |
| KILLARNEY REVOLUTION | | LATE PRE-CAMBRIAN | | |
| | | | | 1.6 BILLION |
| LAURENTIAN REVOLUTION | | EARLY PRE-CAMBRIAN | | |
| | | | | 2.7 BILLION |
| | | PRE-CRUSTAL ROCKS | | |
| | | | | 4.5 BILLION |

MAN APPEARS
FIRST USABLE FOSSILS
OLDEST LIFE RECORD
OLDEST ROCKS
(TRUE TO SCALE)
DURATION OF EARTH TIME.
SEPARATION OF CORE & MANTLE

*Fig. 18*

112

circles much of the way around the earth. One belt of these follows the western edges of North and South America and continues on the western side of the Pacific. Another belt goes almost at right angles from the East Indies through the Himalayas and the Alps.

It is interesting to note in these historical maps that there is some tendency for the continents to have grown outward from central nuclei by the addition of belts of mountains at the margins. This is not always true. Some of the oldest mountains terminate sharply against an ocean. Not enough work has been completed to answer all the questions yet, and there are many interesting ideas to be explored. For example, one of the boldest ideas put forth has been supported by a group of European geologists led originally by Alfred Wegener. This hypothesis has the continents drifting apart from a central single land mass.

## Life and Evolution

What about life on earth? How long did it take for evolution as we think of it to produce a man from the simplest forms of life?

Remains (imprints of living forms) described by Elso S. Barghoorn and his co-workers at Har-

---

Fig. 18. (opposite) *The continents are made up of ancient, eroded mountain belts, like logs in a raft. These developed at different times in earth history, from a measured 2600 million years ago to the present. The geological periods are related to these great mountain-building events.*

113

vard occur in flint nodules in the Gunflint formation in Michigan. They represent early plant remains, similar to algae or fungi, and have been beautifully preserved in solid masses of silica. Age measurements of the enclosing rocks indicate that they are not less than 1600 million years old. Plate IV shows what they look like under the microscope.

Vast expanses of time were needed for the life on the earth to grow more complex. For a billion years it evolved from these primitive forms, leaving very little record of what was happening. The time from 1600 million years ago to 700 million years ago is almost a blank. Although the plants and animals were developing into responsive, specialized, and well-adapted organisms, they had not yet acquired the ability to grow a hard protective shell or structural hard parts that would be preserved as a fossil record of their existence. But, indirectly, the soft algal colonies of very early times left a fossil record by their effect on the precipitation of calcium carbonate in sea water. This was not sufficient, however, to permit detailed studies of the organisms themselves.

About 600 million years ago the first important fossil record started. Prominent types of organisms found in a marine environment at this time are illustrated in Plate V.

We can best understand the streams and the principal dates in the evolution of living things by following them in the diagrams given in Fig. 19. The search, the classification, the reconstruction of the environment, as parts of the prodigious

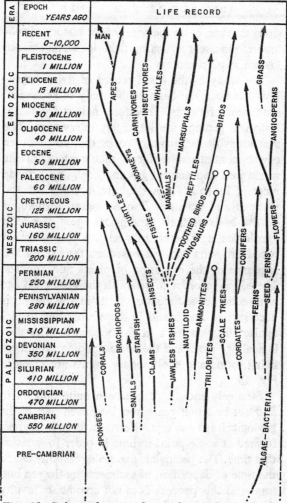

| ERA | EPOCH / YEARS AGO | LIFE RECORD |
|---|---|---|
| CENOZOIC | RECENT 0-10,000 | |
| | PLEISTOCENE 1 MILLION | |
| | PLIOCENE 15 MILLION | |
| | MIOCENE 30 MILLION | |
| | OLIGOCENE 40 MILLION | |
| | EOCENE 50 MILLION | |
| | PALEOCENE 60 MILLION | |
| MESOZOIC | CRETACEOUS 125 MILLION | |
| | JURASSIC 160 MILLION | |
| | TRIASSIC 200 MILLION | |
| PALEOZOIC | PERMIAN 250 MILLION | |
| | PENNSYLVANIAN 280 MILLION | |
| | MISSISSIPPIAN 310 MILLION | |
| | DEVONIAN 350 MILLION | |
| | SILURIAN 410 MILLION | |
| | ORDOVICIAN 470 MILLION | |
| | CAMBRIAN 550 MILLION | |
| | PRE-CAMBRIAN | |

*Fig. 19. Paleontology is the study of the evolution of living things. This diagram summarizes the main courses of development.*

115

study that makes up the pages of the history of evolution, represent a major field in earth science known as paleontology.

The study of the fossil record of living organisms has been the basis of the study of geological sequences of events. Therefore, paleontology and historical geology are intimately interwoven. The vast areas of sedimentary rocks that cover much of the continental surfaces, that contain all the world's oil and natural gas and coal, and provide most of its soils for agriculture, are studied and mapped and classified into geological formations on the basis of their "geological age." The term "geological age" refers to a relative time scale not based on years ago but based on the established sequence of the evolution of plants and animals. Therefore, in order to establish the geological age of a section of sedimentary strata in any area, it is necessary to find fossils in these rocks and to recognize their position in the sequence. The geological age scale and the corresponding prominent living forms and evolutionary events are given in Fig. 19.

Strangely enough, it has been quite difficult to find out in actual years how long each part of the geological age scale is. We need to have crystals formed at some point in time in order to measure that time. The geological age scale primarily charts the time of deposition of sediments on the sea bottom, and this process does not commonly involve the growth of stable crystals. (There are some, however.) Therefore, the time measurements must be made with different rocks, such as have cooled from hot magmas, that lie between the sedimentary sequences. Considerable uncertainty still exists, but the dates attached to the "geological ages" are probably correct within reasonable limits.

You will naturally wonder why the geological age scale ceases to list specific events earlier than about 600 my. ("my." stands for "million years ago") despite the fact that there were major geological events occurring in continental areas for 2000 million years prior to that. The reason is that the fossil record stops at 600 my., leaving the geologist no relative age scale to read. Therefore, the absolute measurement of geologic time by the methods described in Chapter IV must be brought into use to cover the 85 per cent of earth history for which fossils do not exist.

## The Ice Ages

If it were possible to turn backward in time a short way—"short," that is, on the geologic scale —what would we see?

The most striking spectacle of all would be an extraordinary sea of barren ice spreading down from Greenland and the Arctic Ocean and gradually covering Canada and the northern part of the United States until it stood almost a mile thick with a front extending from New York to Oregon, as in Fig. 20. Thick forests of boreal spruce and pine covered the rest of the country to the south. Around the coastlines were broad strips of land, now under water; the level of the ocean had dropped 450 feet below present-day sea level.

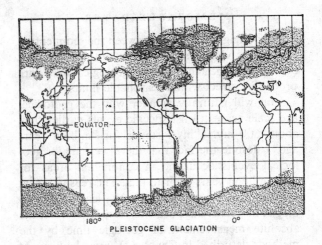

PLEISTOCENE GLACIATION

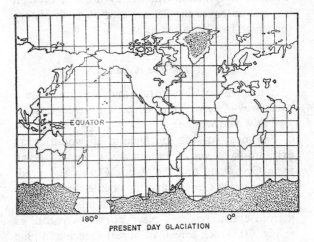

PRESENT DAY GLACIATION

*Fig. 20. Approximately 20,000 years ago the Arctic ice cap extended with great thickness halfway down our continent. This was only "yesterday" in geologic time.*

118

It is hard to believe, but this happened only 10,000 to 15,000 years ago, and it happened at the same time in the northern part of the Eastern Hemisphere. Climates the world over were greatly changed. Types of vegetation and animal life migrated to regions where they do not now exist. The ice is probably still receding today (see Fig. 20), and we see evidence that the sea level in general is still rising.

Going back further in time, we would find the ice receding and then advancing again several times. This continued until we get back to about 600,000 years; there is no evidence of continental glaciation from then until about 200 million years ago. The last three quarters of a million years, or rather this period of ice advance and recession, is referred to as the Pleistocene epoch, or familiarly as the "Ice Ages." It is a very recent time in earth history and is intimately linked with the earlier archaeological history of man.

The measurement of the time of these ice ages is an interesting demonstration of ingenuity in the face of difficult conditions. It has not been possible to use argon/potassium or similar radioactivity methods for this span of time, both because of the very small amounts of radiogenic elements formed compared to contaminations and because of the lack of formation of any crystalline substance that is caused by, or associated with, the ice advances. After a few preliminary remarks on the nature of the problem, let us follow the roundabout way in which this dating problem is being accomplished.

The evidence for the several ice advances appears in the form of planed surfaces of rock that once were hills, great piles of debris, such as boulders, sand, gravel, and wind-blown dust. As the ice overrode the land and receded again, it changed the drainage patterns by damming up valleys and basins, and formed the many lakes in the northern part of the continent, including the Great Lakes.

Glacial geologists have struggled with the history of these advances and have worked out four main advances which agree independently on the North American continent and in Europe. These are spaced by interglacial periods in which warm climates extended deep into the present polar regions. In these times the evolution of present-day man was rapid, so that the dating of the glacial events is intimately associated with anthropological dating and the unfolding of one of the most interesting studies in science—man's emergence to a position of dominance.

How is this dating accomplished? Rough estimates from the counting of annual layering in sedimentary clay deposits, thickness of soil cover, volcanic events, and evidence of man's advancing culture were assembled gradually but still were inadequate. It was not until three independent lines of study were brought together within the last few years that it has been possible to make much progress on this problem.

First, the study of ocean-bottom sediments was made possible by the development of coring devices that could be lowered to the ocean floors to obtain core samples of the bottom muds to depths of several tens of feet. The "piston corer," developed by B. Kullenberg for Hans Petterson's Swedish Deep-Sea Expedition of 1947, brought up cylindrical samples of the bottom material more than sixty feet in length, some of which represent a time span of more than a million years. These cores provide valuable records of recent geological history, as they are often undisturbed by the forces of erosion that act on continental materials. Of particular interest is the fact that the sediments contain the remains of single celled organisms called *foraminifera* which live in the column of ocean water above.

The various species of foraminifera live either on the deep ocean bottom or near the surface, and the skeletons of the different types can be selected from the oozes that make up the cores. These organisms have also evolved with time, and it is possible for paleontologists to use extinct species as a means of determining the geological period in which the organism lived. By studying living forms in different temperature environments, it has also been possible to distinguish the species that live in cold water on the bottom of the oceans from those that inhabit the warmer surface waters.

The next link in the chain of events leading to the answers we need was the development of a "thermometer" that would tell the temperature of the ocean water in the past. This seemingly impossible feat was accomplished by Harold C. Urey, a Nobel prize winner, and his associates at the University of Chicago. Urey predicted that in processes such as evaporation, precipitation, or biological secretion there would be a sufficient fractionation of the isotopes of elements of low mass to be detectable by very refined means of isotopic analysis. In other words, he found that when a glass of water evaporates, the three isotopes of oxygen (oxygen 16, 17, and 18) will not have the same relative abundance in the vapor as in the remaining liquid. Specifically, the lightest isotope, oxygen 16, will be more abundant in the vapor than in the remaining liquid. Similarly, the oxygen of the carbonate, or shell, would be slightly enriched in the heavier oxygen 18 isotope. Of particular interest, however, was the fact that the temperature of the water would affect the amount of the slight enrichment process. Urey predicted that a difference of one degree centigrade in water temperature would produce a difference of only .02 of 1 per cent in the ratio of oxygen 18 to oxygen 16 in the carbonate.

The extraordinary difficulty of detecting and measuring such small differences was finally overcome by redesigning the mass spectrometer to improve its sensitivity by a factor of ten. Finally, after intensive study and outstanding instrument design, it is now possible to measure relative temperatures in carbonate shell fragments to 1°C, if the fragments developed from water that contained the same ratio of oxygen 18 to oxygen 16. Actually, ocean waters differ considerably in this ratio, so that the job of measuring ancient ocean temperatures has been fraught with difficulties; it still is not often possible to place the temperature variations accurately on an absolute scale. But the "thermometer" is now reasonably well established and in many cases can be used for measuring past temperatures with a fair degree of reliability.

Finally, the development of the carbon 14 method of dating provided the last step leading to the solution of our problem. Although carbon 14 age measurement will not extend back in time beyond the last glacial advance, it has provided the yardstick for extrapolating time back through the entire glacial epoch.

Relying on the preliminary studies of Epstein, Craig, and others, an investigation by Cesare Emiliani and his associates at the University of Chicago has finally brought together all these developments in the following demonstration. It was reasonable to assume that the ice advances would be caused by, or reflected in, changes of temperatures of the main bodies of oceanic waters. How true this correlation would be in detail was to be discovered by using the carbonate "thermometer" of Urey on foraminifera of the kind that lived near the ocean surface.

It was first necessary to establish the relative temperature at which various species of foraminifera live at the present time. The species living within a few hundred feet of the surface of ocean waters were selected as giving best indications of detailed variations in climatic history, since the changes in deep ocean water are largely controlled by deep running currents bringing water from thousands of miles away. Then the isotopic ratios were determined in the same species in a number of cores taken thousands of miles apart. By measuring the same species of

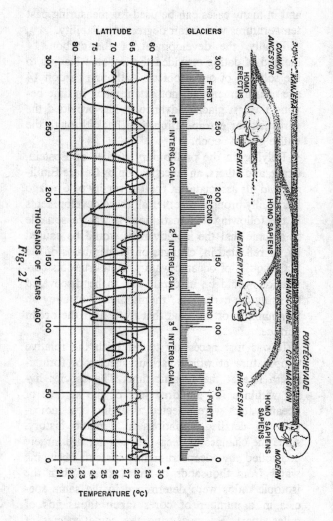

Fig. 21

foraminifera down the length of the core, representing progressively older periods of time, it was possible to observe the variation in temperature of the water of the ocean overlying that core in a series of climatic fluctuations. The picture of these climatic fluctuations appeared to be similar in cores taken from the Mediterranean, Caribbean, Equatorial Atlantic, and North Atlantic areas.

To establish a yardstick of time measurement in these investigations, Hans Suess and Myer Rubin of the Radiocarbon Laboratory of the U. S. Geological Survey determined the ages of the topmost foraminiferal remains by carbon 14 analysis and discovered that the most recent low point in ocean surface temperature came about 18,000 years ago. This agreed with other measurements as being the time of the maximum development of the last

---

*Fig. 21. (opposite) The earth's climate in the last 300,000 years may have undergone the changes indicated in these two curves, which represent an attempt to date the Ice Ages. They correlate geological evidence of ice advance and recession with ocean temperatures and with calculations of variations in the sun's heat. The solid line shows the ocean temperatures derived from study of six cores obtained in deep-sea bottom borings. The gray line charts changes of heat received from the sun in northern latitudes. According to the calculations, latitude 65 receives as much heat from the sun in summer today as latitude 75 did 25,000 years ago. The similarity of the results of the two approaches to dating the Ice Ages is apparent in the curves. The skulls are associated with the epochs indicated on the chart. (After Cesare Emiliani)*

glacial advance in North America. A composite curve showing Emiliani's findings in six deep-sea cores in Fig. 21 shows the climatic fluctuations during the glacial epoch.

There was rather extraordinary agreement between this composite curve and the predictions of the Serbian physicist, Milutin Milankovitch, who in the 1920s worked out the fluctuations in the reception of heat from the sun at various latitudes that might occur as a result of changes in the earth's orbit and in its axis of rotation. Emiliani's comparison of his oceanic temperature variations with these calculated variations in summer solar radiation received at latitude 65°N is also shown on the figure. The curve showing the variation in the sun's heat received in the northern latitudes is expressed as apparent shifts in latitude and shows an apparent 20° maximum equivalent latitude variation in a cycle of approximately 40,000 years.

Although this correlation is not yet proved, because direct radioactive age measurements cannot be extended beyond the most recent glacial advance, it is sufficiently good to constitute a provisional time scale for the entire glacial epoch and is now being tentatively accepted over other methods. If the correlation continues to hold in the future, even the fine details of the glacial advances will be known in the time scale, and a yardstick will be provided for the interpretation of the evolution of man from his predecessors.

No one knows for sure the cause of these ice advances, and we do not know whether another one is due in the next few tens of thousands of years. The balance between precipitation and evaporation in the polar regions is critically dependent upon atmospheric circulations and mean temperatures, and there are many possible ways in which this balance might be shifted.

Geologists have worked out the succession of ice advances and recessions independently on the North American continent and in Europe, and the results seem to agree reasonably well. Four main advances have been recognized, together with interglacial periods in which the warm climates extended even deeper into the polar regions than today. Fig. 21 shows a general pattern of these events.

## Man's Early History

Man appeared on earth only yesterday in geologic time. More than 99.9 per cent of earth time had passed before the first evidences of the special aptitudes of man appeared in the fossil records. The particular study of man's development as a distinct species is part of the branch of science known as anthropology. The study of man's cultural history is known as archaeology.

Archaeological history, like geological history, has therefore been developed on a relative time scale, i.e., one based on something which has continuously changed with time. In geology we have seen that this changing "something" was the universal evolution of living organisms. In archaeology it has been the combined evolutionary changes in man's physical frame and the nature and products of his culture.

It was a great day for archaeologists when W. F. Libby and his associates discovered and developed the use of radioactive carbon 14 for the measurement of very young ages. This method, described in a preceding chapter, is based on a 5600-year half-life so that it is particularly useful in the range of 500 to 30,000 years ago.

Nowadays carbon 14 measurements by the hundreds are illuminating man's cultural history. It could not have been more fortunate that carbon 14 happens to be formed by neutrons in the earth's atmosphere: carbon enters into so much of the matter connected with living things. Shells, flesh, hair, wood, peat—organic remains of all kinds—all contain carbon and can be used for age dating by this method.

For an example of the way in which an archaeological site reveals a carbon 14 record of man's cultural history and evolution, let us look at a recent excavation in the hills of northern Iraq. The site is the huge Shanidar Cave, now occupied by a clan of Kurdish goatherds and their animals. The exciting story unfolded in the course of the excavation has been told by Ralph S. Solecki, who initiated and supervised the work for the Iraqi Directorate-General of Antiquities and the Smithsonian Institution.

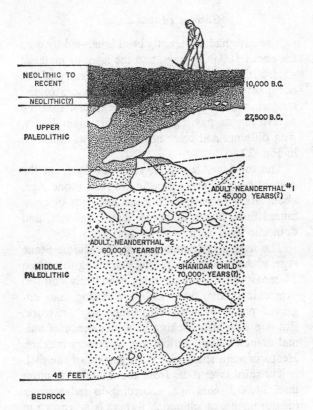

NEOLITHIC TO RECENT

NEOLITHIC(?)

UPPER PALEOLITHIC

MIDDLE PALEOLITHIC

10,000 B.C.

27,500 B.C.

ADULT NEANDERTHAL #1
45,000 YEARS(?)

ADULT NEANDERTHAL #2
60,000 YEARS(?)

SHANIDAR CHILD
70,000 YEARS(?)

45 FEET

BEDROCK

*Fig. 22. Archaeological excavations in the great Shanidar Cave in northern Iraq show a record of late and early man going back about 100,000 years. The record has been dated by carbon 14 in the upper layers. (After Dr. Ralph Solecki)*

129

## Carbon 14 and a Cave

The cave had apparently been inhabited by man for about 100,000 years, and the layer of material 45 feet thick on the floor reveals a history of the occupants going back to the predecessors of the modern species of man. Dr. Solecki describes the excavation as passing through four main layers, with different soil color and artifacts, as indicated in Fig. 22.

The top layer is a greasy soil dating from the present back to some time in the New Stone Age, about 7000 years ago. In it are ash beds of communal fires, bones of domesticated animals, and domestic tools such as stone mortars.

The next layer dates back to the Middle Stone Age, about 12,000 years ago, according to carbon 14 measurements. It contains beautifully chipped projectile points, bone awls for sewing, and engraved pieces of slate suggesting early art work. But conspicuously lacking is any evidence of animal domestication, agriculture, or pottery making. Heaps of snail shells give a suggestion of the diet.

The third layer dates from about 29,000 to more than 34,000 years ago, according to radiocarbon measurements on charcoal. (There is no record in the interval between 17,000 and 29,000 years ago.) This culture is Old Stone Age. Many flint woodworking tools and scraping tools were found, similar to others in the late Paleolithic culture in Europe.

In these three top layers the peoples were all

130

presumably Homo sapiens. In the bottom layer, extending from a depth of 16 feet to bedrock at 45 feet, were found the remains of extinct Homo neanderthalensis, or Neanderthal man. Not only were his crude tools discovered but also three skeletons.

In the bottom layer there is an 8-foot section with an especially heavy concentration of remains of fires, suggesting continuous occupation of the cave through a cold period. There also is a layer of stalagmitic lime, proof that the period was unusually wet—the only one in the history of the cave. The bottom layer may represent time back to 100,000 years ago.

Unfortunately, the radiocarbon of those times is so nearly gone that it can no longer be used effectively. Although it may eventually be possible to extend the method by enrichment of the carbon isotope before measurement, problems of background radiation and of contamination have been insuperable. Dating of most of the Pleistocene and the period of evolution of modern man is currently beyond the reach of the technique. Thus, at present, the dating must be done by geological correlations.

This illustration of the record of man's progress again shows the painfully slow changes in early culture as millennium after millennium passed without variation in rudimentary tools, followed by the almost sudden burst of development that occurred during the last advance and recession of the ice.

Correlation of all present-day knowledge of the

evolution of Man with Emiliani's time scale suggests the progression chronologically charted in this table:

| | |
|---|---|
| Neanderthal man | Became extinct about 50,000 years ago |
| Fontéchavade remains | About 100,000 years ago |
| Swanscombe skull bones | About 125,000 years ago |
| Pithecanthropus, Sinanthropus, and Atlanthropus | About 200,000 years ago |
| South African man-apes | 200,000 to 400,000 years ago |

Thus, if Emiliani is correct, the evolution from Pithecanthropus to Swanscombe man took about 3000 generations, and from Swanscombe to Modern man, only about 1000. The unsuccessful Neanderthal man lasted only about 2000 generations.

The development of man's particular faculties was rapid in this last glacial period. Changes of environment were severe, and his existence depended upon rapid adaptability to climate, food supply, and defense. He survived through craftiness, taking advantage of natural aids wherever possible. Constructing shelters against climate, developing weapons and utensils for food and defense, using fire for his comfort, he gradually became less the nomadic predator and more a member of tribal groups with community of interest. This directed him to the beginnings of agriculture and domestication of animals, and the start of modern civilization.

The use of fire goes far back in man's history. Certainly Neanderthal man used it. Probably the earliest date that can be safely assumed for the use of fire is that of Sinanthropus, or Peking man, the primitive human being who had a brain only two thirds the size of modern man's. Peking man, according to geological estimates, lived about 250,000 years ago. The startling discovery in South Africa a few years ago of Australopithecus prometheus, in deposits showing traces of burned bone, suggested the use of fire by a sub-human creature with a man-ape cranium, but there is insufficient supporting evidence to establish this as fact.

How badly we need an age-dating method that bridges the gap between carbon 14 and the conventional geological methods! The major problem is lack of materials that were formed at the time of the event and preserved intact ever since. Unfortunately, the best material, carbon in the charred wood, which is so commonly found in archaeological sites and which is so useful in carbon 14 dating, is no longer of use in these earlier times.

The answer may have to come from a complete dating of past climates and ice ages and better geological correlation with these.

CHAPTER VI

# The Earth's Beginnings

We cannot consider the origin of the earth without bringing in the origin of the solar system. Because the solar system contains a wide range of elements, we must find some way to account for these, and in so doing we should consider other stars besides our sun. It seems probable that many, if not all, the elements except hydrogen are produced in the interior of stars by thermonuclear reactions. These reactions require exceedingly high temperatures, which, according to calculations, exist only in the interiors of certain stars at a particular time in their history. (Estimates of the temperatures needed are charted in Fig. 23.)

In the center of the sun the pressure is so high that ordinary elements would not be able to maintain the electron shells we are familiar with on the surface of the earth. If its temperature were not exceedingly high, the sun would be a much smaller and denser body than it is known to be.

134

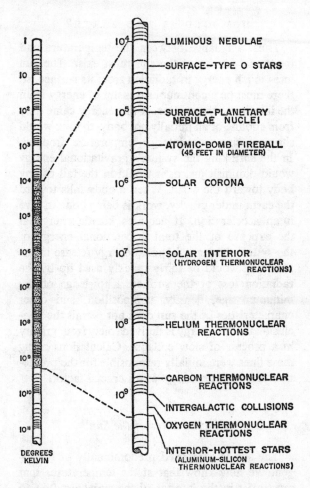

Fig. 23. A popular current hypothesis states that the elements other than hydrogen in the universe were formed by thermonuclear reactions in the center of hot stars. Estimated temperatures for such reactions exceed $10^9$ degrees C.

135

Thus, a balance between the temperature and the mean density of the sun must exist. The sun loses much energy in radiation from its surface, so there must be a continuous transfer of energy from the interior to the surface. If this energy came only from cooling of an initially hot body, the sun would shrink in size as its interior temperature dropped. In the shrinking the system's gravitational energy would diminish just as it would in the fall of any body toward the earth. When a body falls toward the earth under gravity, work is being done to give it an acceleration. It acquires kinetic energy at the expense of the total gravitational energy of the system. The gravitational energy release in the shrinking would be fairly quickly used up by the radiation loss at the surface. Knowledge of the minimum age, density, composition, and other characteristics of the sun does not permit the conclusion that its energy is derived solely from gravity in a process of slow collapse. Calculations along these lines were initially responsible for the prediction that nuclear energy was a crucial factor.

## Nuclear Fires of the Sun

The sun is composed predominantly of hydrogen. We now know that at the temperatures that must exist in the interior of the sun (Fig. 24) to maintain its relatively low density, hydrogen nuclei will react with each other and form helium, with a great release of energy. In other words, hydrogen is being burned up leaving helium as an "ash." The

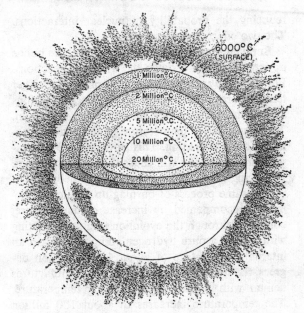

*Fig. 24. The temperature distribution and energy production in the sun have been a subject of much interest, as they were important in predictions relating to thermonuclear reactions. Present estimates of temperature are shown above.*

temperature of this transformation is of the order $10^7$ degrees centigrade.

Why does not the hydrogen react explosively and blow the sun apart, as in a hydrogen-bomb detonation? The answer is that the reaction is automatically controlled by expansion of the sun. The release of thermonuclear energy raises the internal pressure and causes the sun to expand, thereby

137

reducing the probability of nuclear interactions. The two effects are in balance.

But it can easily be seen that as the helium builds up it will also slow down the hydrogen reactions. For the same energy release there must be a higher temperature to increase the probability of reactions among those atoms not yet converted to helium. Thus the interior temperature of the sun is probably slowly increasing. This slow increase of temperature eventually will cause the helium to react, producing carbon 12, oxygen 16, and neon 20.

The same process of burning up the lighter elements, accompanied by increase of temperature, continues through the evolution of a star. Starting with almost a pure hydrogen composition and an internal temperature of about 12–15 million degrees, the star burns its hydrogen and manufactures helium with a gradually increasing temperature. The temperature increases to about 100 million degrees by the time the hydrogen has half gone and the helium starts to burn. Just as the hydrogen becomes exhausted, so eventually does the helium. By the time the temperature has risen to about 600 million degrees, carbon is being converted to sodium, magnesium, and neon. Further increases in temperature cause reactions that produce aluminum, silicon, sulphur, phosphorous, chlorine, argon, potassium, and calcium. Finally, at about 2000 million degrees these elements are converted into what is known as the iron group—titanium, chromium, manganese, iron, cobalt, nickel, copper, and zinc.

Probably the entire star would not become converted to these elements because the temperature would decrease toward the surface. How the elements, from hydrogen to the heaviest, are distributed between the surface and center is difficult to estimate, but in the sun it appears likely that at least the outer half of the radius is made up of fairly homogeneous material.

In this process of evolution, well summarized recently by the British mathematician Fred Hoyle, William A. Fowler, of California Institute of Technology, and others, the star's interior temperature will increase and the star will become extremely expanded and more luminous. Also its surface temperature will increase. It is calculated that the sun, which is fairly youthful in this stellar life span, eventually will become 1000 times more luminous and have a radius 100 times greater than at present. Finally, it will shrink to one twentieth of its present size as the nuclear fires within die out, and it will become the type of star known as a "white dwarf."

This hypothesis of stellar evolution is supported by observations of the relationship between luminosity (brightness) and surface temperature (color) of numbers of stars. Fig. 25 shows, for example, a study made by H. C. Arp and A. R. Sandage of the Mount Wilson and Mount Palomar observatories, in which this relationship is plotted for a number of stars belonging to the globular cluster Messier 3. All the stars plotted, it is believed, have a mass fairly similar to that of the sun.

The plot makes a definite pattern. The hypothesis holds that a star will start its evolution close to

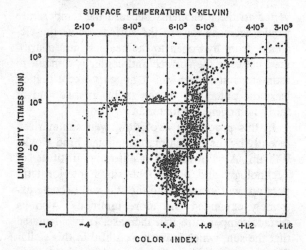

*Fig. 25. The evolution of stars is an exciting new subject. An important piece of evidence in a chain of reasoning is this compilation of the relationship between color and brightness in a number of stars in a globular cluster. (After H. C. Arp and A. R. Sandage of the Mt. Wilson and Mt. Palomar observatories)*

the bottom. Its luminosity (dependent on size and temperature) increases, until it reaches the "giant" branch at the upper right. Finally, the luminosity decreases slightly, but the surface temperature increases greatly, and the star follows the trend shown in the horizontal branch at the upper left. In this stage the light that is emitted shifts farther to the ultraviolet (indicating a higher surface temperature). This ultraviolet is strongly absorbed by interstellar gas and our own atmosphere, and in time the star disappears from our view.

## Origin of the Elements

After a brief look at these hypotheses, let us now return to the origin of the elements that make up our solar system. It is not yet known for certain whether stars considerably larger than our sun might be capable of making the heavier elements, but it is an attractive thought. If true, it seems likely that the elements would have been dispersed into space within our galaxy through the explosions of supernovae. These exploding stars have been estimated to have been frequent enough to account for the small amount of material heavier than helium in the basic composition of our galaxy.

One of the problems in this hypothesis is the apparent uniformity of composition of the universe. The American astronomer Armin J. Deutsch has discussed the evidence from solar, stellar, and interstellar absorption spectra (the analysis of light from the sun and stars), indicating the extreme uniformity of their composition if adequate corrections are made for temperature and other effects. The observed "cosmic" abundances of the elements are indicated in Fig. 26.

Hydrogen greatly predominates, with helium next, and all the rest of the elements make up less than 1 per cent of the total. We are very interested in the relative abundances of the elements heavier than helium, because, as we shall see from the relationship between uranium and lead isotopes on the earth and in meteorites, the elements in this part of our galaxy are no older than a few billion years.

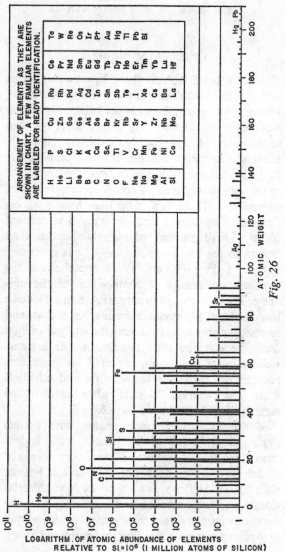

ARRANGEMENT OF ELEMENTS AS THEY ARE
SHOWN IN CHART. A FEW FAMILIAR ELEMENTS
ARE LABELED FOR READY IDENTIFICATION.

| H | | | | | | He |
|---|---|---|---|---|---|---|
| Li | Be | B | C | N | O | F |
| Ne | Na | Mg | Al | Si | P | S |
| Cl | K | A | Ca | Sc | Ti | V |
| Cr | Mn | Fe | Ni | Co | Cu | Zn |
| Ga | Ge | As | Se | Br | Kr | Rb |
| Sr | Y | Zr | Nb | Mo | Ru | Rh |
| Pd | Ag | Cd | In | Sn | Sb | Te |
| I | Xe | Cs | Ba | La | Ce | Pr |
| Nd | Sm | Eu | Gd | Tb | Dy | Ho |
| Er | Tm | Yb | Lu | Hf | Ta | W |
| Re | Os | Ir | Pt | Au | Hg | Tl |
| Pb | Bi | | | | | |

*Fig. 26*

LOGARITHM OF ATOMIC ABUNDANCE OF ELEMENTS
RELATIVE TO Si = 10⁶ (I MILLION ATOMS OF SILICON)

142

Does this mean that the entire universe is only a few billion years old? Or just our galaxy?

Turning to our planetary system, we have the following evidences and problems. The densities (calculated for zero pressure) compared with the size of the bodies near the sun are as follows:

|  | Mass (compared to earth) | Calculated Mean Density (zero pressure) |
|---|---|---|
| Moon | 0.012 | 3.3 |
| Mercury | 0.054 | ? |
| Mars | 0.108 | 3.6 – 4.0 |
| Venus | 0.81 | 4.3 – 4.7 |
| Earth | 1.0 | 4.4 |

This suggests that the larger the body, the heavier is the material it is made of. But is this due to a greater abundance of heavier elements brought together at the start? It is difficult to believe that the moon and earth were not originally of the same composition, since they presumably condensed from the same cloud of gas and dust.

It has often been stated that the earth's core of iron and the moon's lack of it cause the difference.

*Fig. 26. (opposite) If the elements were formed in thermonuclear reactions at the center of hot stars, their relative abundances should be predictable to some extent. For this we have to know the neutron capture cross sections (the probability of interaction between a given nucleus and an incident neutron) and the stability of the various nuclei. It has been of interest to correlate the observed abundances in nature with the predicted ones. The agreement is quite good.*

But why would the abundance of iron in the earth differ from that in the moon? G. J. F. MacDonald has recently proposed that the relative abundance of the heavier elements in the planetary bodies was initially the same. In the earth, however, the pressure was not sufficient to strip electrons from atoms, as in the sun, but was sufficient to result in a separation of oxygen from iron and possibly silicon in the central region. The lower melting temperature of the pure metal phases permitted fusion.

It is still bothersome, however, that the ratio of iron to silicon in the sun appears to be much less than in the earth or meteorites. As the temperature of the interior of the sun is definitely not sufficient to be manufacturing silicon, magnesium, or iron, the sun inherited its supply of these elements from a source, presumably (but not necessarily) the same as that of the planetary bodies. This source probably developed early in the history of our galaxy as a cloud of dust or gas of cosmic composition, and was one of a large number of such clouds condensing into stars similar to the sun.

## From a Cloud of Dust

In the last few years G. P. Kuiper, developing and modifying the ideas of C. F. von Weizäcker, has contributed much to an hypothesis of wide acceptability on the origin of the solar system. The central part of the cloud condensed to form the sun. The rotating disc of dust and gas which formed quickly from the nebula surrounding the sun broke up into eddies of irregular size and arrangement within any radial zone, but the size of the eddies increased generally toward the outer zones. The eddies in any zone coalesced as they met each other, until finally there remained separate masses of gas at different radii from the sun. These masses of gas and dust have been referred to as "protoplanets."

The theory appears to provide rather nicely for the spacing and total mass of the protoplanets, the distribution of angular momentum in the solar system, and the sense of the rotation of the separate masses.

Among astronomers the formation of the planets from the protoplanets is a lively question at present. Much evidence must be accounted for. In mass the terrestrial planets are very much smaller than the protoplanets were, and very different in composition. The protoearth was a low-temperature gaseous mass about 500 times the planetary mass; the sun had not condensed sufficiently to be hot. The protoearth gas cloud was composed of hydrogen, helium, neon, methane, ammonia, and, possibly, some water vapor. Materials of which the present earth is made were condensed and eventually spiraled into the center. A secondary condensation nucleus formed the moon.

When the sun contracted and became bright, its radiations energized the cold gases of the protoplanets. Individual particles acquired velocities that permitted them to escape from the protoplanet's gravitational field. The escape of these gases, which amounted to 99 per cent of the protoplanet mass, was essentially complete, even of the heavier ones like krypton and xenon. According to Kuiper's estimates, the time required for absorbed solar radiation to cause this removal of gases was probably several hundred million years.

We can follow several lines of evidence in an attempt to form an hypothesis for the early stages of the accretion and development of the earth system. The surface of the moon, which has no atmosphere and has retained its original character since its beginnings, shows great craters and scars, as if struck by large objects in its growth (see Plate VI). The origin of the iron and stony meteorites must be accounted for, as well as the apparent difference in iron content of the earth and moon.

The total amount of energy released by the infall of material to form the earth was not sufficient to have raised its temperature to its probable present value. Continuous radiative loss of heat outward also reduced the retained energy. So, depending on the rate at which the earth accreted, it could have been formed at various temperatures probably not exceeding about 2000° C.

Harold C. Urey, basing his arguments on the lack of concentration of volatile elements at the surface, has suggested that the temperature of the earth's surface in its final stages of accumulation probably was not higher than 200° C. Earlier theories, which conflicted with each other, were largely based on the assumption that the earth was molten in the beginning. Today the general belief is that initially the earth was fairly cold. Radioactivity, then 15 times more intense than it is now, and energy that may have been released when the iron core material settled to the center raised the temperature to a level at which the earth was essentially molten. It was at this time that the gravitational separation of core, mantle, and probably some of the materials of the ocean and atmosphere occurred.

Sir Harold Jeffries has developed the theory of a cooling molten mantle, crystallizing from the base upward, with heat lost first by convection in the liquid and later by conduction in the solid. His outstanding treatise has encountered the objection that the earth probably formed as a solid body, but the theory could apply if the mantle ever became molten at a later time. The crystallization from the base outward would result from the fact that a mixture of the oxides of the rock-forming elements solidifies over a range of temperature; the more stable forms would crystallize at a higher temperature and sink in liquid of lower density.

## The Answer

Let us now return to the measurement of the age of the earth and solar system. So far, we have discussed age measurements that have depended upon parent-daughter relationships in a mineral that contains the parent element. Another interesting part of the field of age measurement relates to the slow change of the isotopic abundance in elements in which one or more isotopes are radioactive and breaking down, or radiogenic and building up.

147

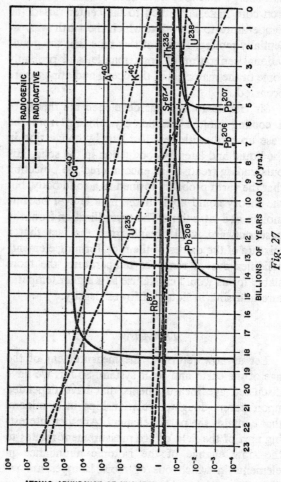

*Fig. 27*

ATOMIC ABUNDANCE OF NUCLIDES
RELATIVE TO 10⁶ ATOMS OF SILICON

148

If we know the half-life of a radioactive isotope, we can calculate the change of its abundance relative to the stable isotopes of the same element and see what happens as we go back in time. In Fig. 27 curves are shown for the changing abundances of potassium 40, rubidium 87, uranium 235 and uranium 238, and thorium 232.

As an example of radiogenic isotopes building up with time, let us look at the element lead. Lead 204 has no known long-lived parent so, presumably, has not changed in abundance significantly in the last few billion years. Lead 206, 207, and 208 are increasing at different rates because they are formed from the breakdown of $U^{238}$, $U^{235}$, and $Th^{232}$, respectively. If we now calculate the change in the ratios 206, 207, and 208 with time going backward, we get curves as shown in Fig. 27.

In these curves we have assumed average estimated values for the abundances of uranium, tho-

---

*Fig. 27. (opposite) Calcium 40, argon 40, strontium 87, lead 206, 207, and 208 are examples of radiogenic nuclides. They are increasing in abundance with time because they are the decay products of potassium 40, rubidium 87, uranium 238, and thorium 232. This means that, going backward in time, there would be less of them until ultimately there was zero. Because the parent element is decaying exponentially, the drop-off of the daughter to zero appears to be quite a sudden phenomenon on the chart. These zero values set the maximum age for the origin of these pairs of elements. Lead 207 has the lowest maximum age, and this sets the limit for our part of the universe.*

rium, and lead. Other values would change the time scale on these curves but not their shape. In each case there is some limiting time in the past beyond which we cannot go because the radiogenic isotope would be reduced to zero.

For example, in the figure we see that lead 207 is reduced to zero at about 7 billion years ago, which suggests that the association of uranium and lead in at least our solar system could not have existed earlier. As it seems likely that these heavy elements were all created together, the maximum time for the creation of the elements appears to be about 7 billion years ago.

Astrophysical calculations set an age limit for the galaxy's globular clusters at about 6.5 billion years. Being almost certainly younger than these, the sun is estimated to have an upper age limit of about 6 billion years.

Now let us approach the problem from the opposite direction and examine the oldest available material to establish a limit on how *young* the earth could be. The oldest measured rocks are 2.8 billion years old, and we know the earth is older than that. Argon/potassium and strontium/rubidium measurements on stony meteorites show a good maximum age grouping at 4.5 billion years. These objects are made of minerals which could not have been formed under pressures such as exist in our earth's deeper regions, so they must have been derived from the collision of small bodies or else were recrystallized from some other kinds of minerals or represent parts of the near-surface layers of a larger body. Thus, the age of 4.5 billion years simply represents the time at which an event happened, which probably was similar to the separation of the earth's core and mantle.

Using a combination of the lead isotope ratios in iron meteorites and the curves shown in Fig. 27 based on the U/Pb ratio in the earth's upper mantle, we also get an age of 4.5 billion years, which suggests that the major separation of the earth into iron core and stony mantle occurred at approximately the same time as the development of meteorites.

Going back to our thoughts on the early history of the earth, we decided that a considerable length of time was needed for radioactivity to heat the earth from an initially cool beginning. Thus, the commonly quoted age of 4.5 billion years for the earth is a lower limit only, and the true age is probably hundreds of millions of years greater. As you will recall, it probably took several hundred million years to remove the gases from the protoplanets. If we add this period to the time needed to heat the earth enough for separation of mantle and core and the time needed for the sun to condense and become hot, we account for the period between 4.5 billion years ago and 6 billion years ago, the estimated time of the sun's origin.

So, in summary, we see that the elements in our part of the universe, the globular clusters and probably our galaxy formed 6.5 to 7 billion years ago; the sun condensed 6 billion years ago; the protoplanets reduced to planets about 5 billion years ago; the chemical separations within the terrestrial planets and parent bodies of the meteorites occurred 4.5 billion years ago; and formation of a lasting earth crust started 2.8 billion years ago.

How majestic are these broad reaches of time! Looking into an abyss, one senses the gigantic form of the void only in comparison to one's own minute stature. It is almost incomprehensible that only a few billion years ago our galaxy was born in a giant bomb-flash of nuclear energy. What an inspiring picture of the process of creation! But awesome and inspiring as it is to contemplate this mighty spectacle, the true reward is not to be found in whether our calculations are correct, give or take a few million years; it lies in the discoveries, in the advancement of human knowledge and philosophy that are the inevitable products of scientific search for law in nature.

# BIBLIOGRAPHY

1. *Physical Geology* by L. Don Leet and Sheldon Judson, Prentice-Hall, New York
   A good general textbook on physical geology.

2. *The Scientific American* book series, Simon and Schuster, New York
   Represents compilations of articles in this magazine on specific topics. In particular, the volumes: *The Planet Earth, The New Astronomy,* and *The Universe.*

3. *The Earth and Its Atmosphere,* edited by D. R. Bates, Basic Books, Inc., New York
   Contains articles by fifteen scientists on the crust and interior of the earth, its origin and ultimate fate; on the oceans, atmosphere, ionosphere, magnetic storms and cosmic rays.

4. *Nuclear Geology,* edited by H. Faul, John Wiley and Sons, New York
   A symposium on nuclear phenomena in the Earth Sciences, including age measurement, earth heat and radioactivity, instruments and uranium prospecting techniques.

# INDEX